MANGROVES FOR BUILDING RESILIENCE TO CLIMATE CHANGE

A Field Manual

MANGROVES FOR BUILDING RESILIENCE TO CLIMATE CHANGE

A Field Manual

R. N. Mandal, PhD
R. Bar, PhD

Apple Academic Press Inc.
3333 Mistwell Crescent
Oakville, ON L6L 0A2 Canada

Apple Academic Press Inc.
9 Spinnaker Way
Waretown, NJ 08758 USA

© 2019 by Apple Academic Press, Inc.

First issued in paperback 2021

Exclusive worldwide distribution by CRC Press, a member of Taylor & Francis Group
No claim to original U.S. Government works

ISBN 13: 978-1-77463-406-6 (pbk)
ISBN 13: 978-1-77188-716-8 (hbk)

Library and Archives Canada Cataloguing in Publication

Mandal, R. N., 1964-, author Mangroves for building resilience to climate change : a field manual / R.N. Mandal, PhD, R. Bar, PhD.

Includes bibliographical references and index.
Issued in print and electronic formats.
ISBN 978-1-77188-716-8 (hardcover).--ISBN 978-0-429-48778-1 (PDF)

1. Mangrove forests. I. Bar, R., 1977-, author II. Title.

QK938.M27M36 2018   577.69'8   C2018-904810-7   C2018-904811-5

Library of Congress Cataloging-in-Publication Data

Names: Mandal, R. N., 1964- author. | Bar, R., 1977- author.

Title: Mangroves for building resilience to climate change : a field manual / authors: R.N. Mandal, R. Bar.

Description: Waretown, NJ : Apple Academic Press, 2018. | Includes bibliographical references and index.

Identifiers: LCCN 2018038768 (print) | LCCN 2018039634 (ebook) | ISBN 9780429487781 (ebook) | ISBN 9781771887168 (hardcover : alk. paper)

Subjects: LCSH: Mangrove plants. | Mangrove ecology. | Mangrove conservation. | Mangrove management.

Classification: LCC QK938.M27 (ebook) | LCC QK938.M27 M348 2018 (print) | DDC 583/.96--dc23

LC record available at https://lccn.loc.gov/2018038768

Apple Academic Press also publishes its books in a variety of electronic formats. Some content that appears in print may not be available in electronic format. For information about Apple Academic Press products, visit our website at **www.appleacademicpress.com** and the CRC Press website at **www.crcpress.com**

ABOUT THE AUTHORS

R. N. Mandal, PhD

R. N. Mandal, PhD, is Principal Scientist in the discipline of economic botany at the Indian Council of Agricultural Research, Central Institute of Freshwater Aquaculture (ICAR-CIFA), Rahara, Kolkata, West Bengal, India. Dr. Mandal received a Young Scientist Award for his outstanding work on mangrove ecology and diversity in 2001 from the Indian Science and Environmental Programme, Science Academy, Gorakhpur, India. He has also been selected as a life member by the National Environmental Science Academy, New Delhi, for his work on mangrove ecology and biodiversity. He has published a number of articles in refereed scientific journals as well as two books: *Ecology and Biodiversity of Indian Mangroves* (Vols. I and II); and *Sundarban,* written in the Bengali vernacular language for introducing mangroves to the general public. Dr. Mandal has been a regular visitor to the Indian Sundarban because of his fascination and love of mangroves. Dr. Mandal earned his MSc and PhD degrees in botany from the University of Calcutta, India. His thesis was titled "Morpho-Anatomical Studies of Mangroves and Halophytic Algal Flora of the Sundarbans in West Bengal (India) with Special Reference to Their Ecological Adaptations." He also worked on the formulation of strategic planning for conservation of mangroves for the Indian Sundarban, in particular, in association with Dr. K. R. Naskar. During his PhD work Dr. Mandal visited almost all known mangrove habitats of the east and west coasts of India and keenly observed their adaptive habits from an ecological perspective of coastal regions.

R. Bar, PhD

R. Bar, PhD, is Assistant Professor of botany at Bangabasi Morning College, University of Calcutta, India. He has been a frequent visitor to the Indian Sundarban to educate students on mangroves. He has published a few articles in refereed journals and edited the book *Sundarbans: Issues and Threats* with Dr. K. R. Naskar. Dr. Bar earned his MSc and PhD degrees in botany from the University of Calcutta and the University of Kalyani, West Bengal, India.

CONTENTS

ABBREVIATIONS

AEP	Atlantic-East Pacific
AFLP	amplified fragment length polymorphism
Arab	Arabic
B	Bengali
Bo	Bombay State
BSI	Botanical Survey of India
C	criteria
CL	consolidated land
CR	critically endangered
DA	data assessed
DD	data deficient
DS	downstream
E	English
EN	endangered
GIS	geographic information systems
GMSL	global mean sea level
GPS	Geographical positioning system
Guj	Gujarati
H	Hindi
HT	high tide
IPCC	Inter Governmental Panel on Climate Change
IWP	Indo-West Pacific
Kan	Kanari
LC	least concern
LT	low tide
M	Madras state
Ma	Marathi
Mal	Malayalam
ML	mudflat land
MS	midstream
MSL	mean sea level
MT	medium tide
NT	near threatened

NTFP	non-timber forest product
Odi	Oriya
P	Panjabi
RLC	Red List category
RMD	relative mangrove diversity
S	Sanskrit
SET-MH	surface elevation table – marker horizon
SL	stable land
SLR	sea-level rise
T. S.	transverse section
Tam	Tamil
Tel	Telegu
US	upstream
VN	vernacular name
VU	vulnerable
YP	year published

FOREWORD

Mangroves are a fascinating group of plants that occur on tropical and subtropical shorelines of all continents. Here, they are exposed to daily saltwater flushing, generally low oxygen levels around their roots, high light and temperature conditions, and periodic tropical storms. Despite these harsh conditions, mangroves may form luxuriant forests that have significant economic and environmental value to coastal communities throughout the world. Yet, mangroves are threatened around the world: in some regions, the threat comes from the immediate pressures of subsistence exploitation, but in other regions, the threat arises from a failure of the general community, and particularly of decision-makers, to truly appreciate the value of these coastal communities. The failure to recognize the economic and ecological value of mangroves in terms of direct products, indirect products, and the amenities or 'free services' they provide has resulted in an attitude that these communities are wastelands, that, in turn, can be converted to 'better' usage. In addition, the failure of scientists to undertake appropriate research and then communicate convincingly those findings to the public and the decision-makers must also receive some share of the blame. This field manual seeks to address this very area of concern: it explains in simple terms what scientific research has revealed about the mangrove plants, their environment, and how the plants and the environment interact. It provides an up-to-date account of the mangrove plants, their detailed identification and associations, their specific habitat requirements and adaptations, their reproductive strategies, and the management requirements of these plants to ensure their sustainability and conservation in a period where climate change provides acute challenges to this group of plants. The authors bring unique expertise and the passion to convey the latest information about these fascinating plants, particularly those that occur in India, thereby helping to reduce the threats that currently confront them.

—**Peter Saenger**
Emeritus Professor
Centre for Coastal Management
School of Environment, Science & Engineering
Southern Cross University
P.O. Box 157, Lismore, NSW 2480, Australia

PREFACE

In writing this field manual, I have been inspired by many individuals within the field. I was privileged to work with Dr. K. R. Naskar, an authority on mangroves floral diversity, particularly in the Indian Sundarban. Dr. Naskar supervised me for my PhD degree from the University of Calcutta and aroused my interest in mangroves and their ecosystem. During my early days of PhD work, he guided me meticulously in identifying each and every species growing in the Indian Sundarban and their ecology. We jointly published one book, *Ecology and Biodiversity of Indian Mangroves*, in 1999. After a few years, Dr. Radharaman Bar, co-author of this field guide, contacted Dr. Naskar with the intention of working on mangroves with him. Sadly, Dr. Naskar left for the heavenly abode, leaving indelible impressions in our minds, with his study of mangroves making a significant mark on the field.

The Indian subcontinent, being situated in a tropical region, has large areas occupied by mangroves, though many of us have little idea about these plant communities. Not only are sizable coastal areas covered with mangroves communities, but the world's single largest block of mangroves forest, the Sundarban, is also situated in the Ganges-Brahmaputra-Meghna estuary. Due to the presence of the Royal Bengal tiger, this unique forest has not been easily accessible to others, although fishermen, woodcutters, and locals venture going deep into the forest with serious risk. Our experience shows us that the minds of students, teachers, foresters, researchers, and others are curious about these plant communities. Which plants are mangroves and why? How do they survive growing in such a highly saline region? Very often we encounter common inquiry due to people's interest in mangroves. Hence, we badly need a book on mangroves that will serve to elucidate the general public—and scientists—on mangroves. There are some books that broadly describe the different aspects of mangroves and coastal vegetation. However, no book we have yet found that mentions the morphology of mangroves can act as a field guide and may be handy to carry during field study. Students, teachers, foresters, researchers, and others interested look for a handy reference to guide them in the field and to assist them with identifying the mangroves that form this unique coastal estuarine vegetation.

This book intends to provide an easy method of identifying mangroves and to distinguish one species from another. What is a mangrove and which criteria should mangroves possess? What are the different attributes, along with distinction among major mangroves, mangrove associates, mangrove halophytes, and back mangals? This volume answers these questions. Images and illustrations of individual species are included to guide readers about the visible features of mangroves.

In this effort:

- Chapter 1 introduces some facts about mangroves: definition, the genesis of the term 'mangrove,' early records of mangroves based on fossil, historical, and mythological evidences, categorization of different mangroves, along with habitats.
- Chapter 2 mentions the distribution of mangroves worldwide, with the support of references and a detailed diversity in India in particular, as well as the features of habitats of mangroves and climatic conditions.
- Chapter 3 describes the ecology and environmental conditions, particularly the concept of intertidal zones along estuary positions where tidal flows inundate the mangroves. Mangroves' succession in response to environmental changes is mentioned, along with the association among different mangroves.
- Chapter 4 describes distinct morphological attributes modified to sustain saline conditions that altogether characterize halophytic adaptation. The reproductive phenology of major mangroves is discussed, which may be essential information in view of climate change.
- Chapter 5 details the mangroves with every attribute in a concise but easy-to-follow way. Students will benefit from this guide, which will help them to identify each mangrove clearly in the field. A total of 78 species of intertidal flora, including 32 true mangroves, are briefly described, together with diagnostic features, salient attributes, and illustrations for easy identification.
- Chapter 6 highlights the burning environmental issue of climate change and its impact on mangroves. Climate change may lead to sea-level rise, but how mangroves cope with such an adverse situation is discussed with strategic points for building resilience among mangroves.

- Chapter 7 was not initially conceived in our planning. Later, a sense of guilt surfaced in our mind as to why methods of restoration and conservation should not be mentioned for those who desire to protect mangroves. Prof. Peter Saenger, a renowned mangroves scientist, pointed out our mistake in failing to provide the techniques of how to save mangroves.

This book is made to be a field guide with the intention of informing the general public and others about mangroves, so their essential attributes relevant to field study are mentioned concisely and clearly, with focus given to every aspect on mangroves for the very purpose of their identification.

However important mangroves may be, people know very little about them. Dr. Naskar, our master, tried to popularize mangroves among common people. He did so in an effort to sensitize them to their different values, generated through ecosystem services that immensely benefit coastal people and protect them like motherly care. He used to persuade common people, particularly coastal inhabitants, to make an effort to restore mangroves that are lost forever. His life was fully dedicated to mangroves and related activities, with the establishment of a 'Mangroves Ecological Park'—one among other noble contributions. He was desirous of preparing one guidebook that might have helped a wide range of people to become acquainted with mangroves—a comprehensive but concise manual with figures and facts, but that guidebook did not materialize due to his sudden departure. Surely his endeavor was much more concentrated on mangroves than that of the average person. He came from the geographical zone in which mangroves once existed, now reclaimed. The euphony 'Sundarban,' covering both reclaimed and forested areas, is both familiar and famous to the world, transcending limited geographical boundary. This field guide is the result of Dr. Naskar's consistent inspiration is still with us, despite his earthly departure. We try our best to satisfy researchers; if not, we are responsible for failing the very purpose of this book: to guide researchers in the field, and that failure is comparable to the withering of mangroves due to missing river flow, caused by excess salinity and scarcity of freshwater. Nevertheless, if this book satisfies students, teachers, foresters, researchers, or even guides a single person to know about and protect mangroves, one soul that is Dr. Naskar, we believe, will rest with peace in the heavenly abode.

—R. N. Mandal and R. Bar

ACKNOWLEDGMENTS

The book is based on the PhD thesis "Morpho-anatomical studies of mangroves and halophytic algal flora of the Sundarbans in West Bengal (India) with special reference to their ecological adaptations," written under the supervision of Dr. K. R. Naskar, from the University of Calcutta. The thesis belongs to the senior author and was completed a little less than a quarter century ago, in the early nineties. The senior author is deeply indebted to two individuals: the late Dr. K. R. Naskar, who stimulated authors in mangroves research, and Shri Girin Raftan, a boatman who determined the whereabouts of each mangrove in the inaccessible, dense vegetation of the Sundarban. Shri Raftan even dared to collect a few mangroves from their dense habitats, taking large risks due to the presence of the Royal Bengal tiger. The author realizes that both Dr. Naskar and Shri Raftan seemed made for each other for the greater cause of the Sundarban: when the former would search a for specific flora through early records, the latter could know its whereabouts simply through discussion, due to his vast experience with mangrove forests.

Dr. A. G. Untawale, Joint Director of NIO (National Institute of Oceanography, Goa, India), is duly acknowledged, as he provided literature from the NIO library and accompanied us (the senior author and Dr. Naskar) during our visit to the Goa estuary. The author would like to thank the staff of Botanical Survey of India (BSI), Kolkata, who allowed him to study the herbariums of mangroves. The author also received generous support from anonymous individuals during the visits to different mangroves habitats such as Bhitarkanika mangroves forest, Odisha; Cochin estuary, Kerala; Bombay coast, Maharashtra; as well as the Pondicherry, Andaman, and Nicobar Islands mangroves.

After almost a gap of two decades, the co-author joined the senior author with the intention of preparing this book. Both authors visited mangroves, checked the attributes, observed their ecology, made some notes, and revised them. They also collated the information and observed the local efforts for the conservation of mangroves. During the preparation of this book, we sought support from many experts and others to whom we extend our sincere gratitude:

- Prof. Peter Saenger, Emeritus Professor of Southern Cross University, Australia, and a renowned mangroves expert, reviewed the early draft of this book, suggested some additions for its improvement, and later reviewed the entire manuscript.
- Dr. Chandan Surabhi Das, Assistant Professor, Department of Geography, Barasat Govt. College, West Bengal, reviewed Chapters 3, 6, and 7, and suggested the necessary changes. He took meticulous effort for the major revision of Chapter 6, which has been relevant in the context of global warming and sea level rise.
- Dr. Saikat Naskar, Assistant Professor, Department of Botany, Barasat Govt. College and younger son of Dr. Naskar, reviewed the entire manuscript and suggested the necessary inclusions, including e-taxonomy through the online database (webs) search (www. theplantlist.org, http://www.catalogueoflife.org and http://www. gvif.org), and advised us to update the author citation of *Bruguiera gymnorhiza, Ceriops decandra,* and *Scyphiphora hydrophylacea* (see Chapter 5).
- Dr. Naskar also corrected the identification of *Merope angulata,* which was misidentified as *Atalantia correa,* and provided the image of *Myriostachya whitiana.*
- Dr. Goutham-Bharathi, Research Associate, ICAR-CARI, Andaman & Nicobar Islands, India, with his profound knowledge, not only provided images of *Bruguiera parviflora, Rhizophora stylosa,* and *Sonneratia ovata,* but also revised and corrected many attributes of mangroves that grow exclusively in the Island mangroves forests. His comments helped us revise the record of *Cynometra iripa.*
- Prof. Gour Gopal Maity, Senior Professor, University of Kalyani, corrected few names of mangrove halophytes and back mangals.
- Dr. D. N. GuhaBakshi, former deputy Director of Botanical Survey of India, Kolkata, reviewed Chapter 1.
- Prof. S. C. Santra, Senior Professor, Department of Environmental Science, University of Kalyani, and my teacher, inspired me constantly in pursuing publication on mangroves.
- Dr. Monahar, Scientist, Coastal Agriculture Research Institute, Goa, gave us images of *Sonneratia alba.* Mr. Avijit Jha gave us an image of the Sundarban mangrove forest.

- Mr. Soumaya Naskar, elder son of Dr. Naskar, requested that we write a field book on mangroves that would help many people to engage in mangroves conservation.

In addition, we received generous support and help during the preparation of this book from Smt. Sipra Naskar, wife of the late Dr. Naskar, Smt. Purnima Mandal, Smt. Shawli Bar and Ms. Priya Singh.

The senior author has been instilled with the blessings of revered Swami Sandarshanananda Maharaj, a senior monk in order, Ramakrishna Math and Mission, Nerendrapur, West Bengal, India, for writing this book.

The authors would like to thank Ms. Sandra Jones Sickels, Vice President, Editorial and Marketing, Apple Academic Press Inc., for her kind consent to publish this manuscript as a book, and her colleagues, Mr. Ashish Kumar and Rakesh Kumar (Production).

Our thanks should go to Prof. Dr. Shigeyuki Baba, ISME; Mr. Kelly Hoff, Permission Coordinator, Copyright & Permissions, Wiley; Prof. Brian McGill, Editor in Chief, Global Ecology and Biogeography; Dr. Beth A. Polidoro, IUCN species Programme/SSC/Conservation International Global Marine Species Assessment, Biological Sciences, USA; Dr. P. Selvam, Executive Director, MSS Research Foundation, Chennai, India; and Dr. D. M. Alongi, AIMS, Australia, who generously consented to reprint the following published Tables 2.1–2.4 and 6.1, respectively.

Both authors photographed all of the images in the field, except for a few that have been collected from others through personal communication, as mentioned above, with permission.

—**R. N. Mandal and R. Bar**

CHAPTER 1

MANGROVE AND ITS HISTORY

1.1   INTRODUCTION

Mangroves—the intertidal forest—struggle every moment against hungry tides as they straddle at the interface zone of the land and the sea. Despite putting as much as effort relentlessly to sustain in fragility, mangroves, however, never deny of displaying their beauty that may be expressed in a line of the poetic verse, 'The woods are lovely, dark and deep' – to keep everyone mesmerized. As evergreen as they look pristine, mangroves form a definite stand of the sheltered coastal vegetation in the tropical regions of the world. This intertidal forest continues to sacrifice in protecting people living in such fragile zones against frequent natural calamities. They bathe twice a day with water of tidal rise and expose to sunlight – a rare occurrence never seen in any vegetation of the planet Earth. Everybody knows roots of a plant to enter into earth and stem to air, but mangroves ignore the common rule, deviating the usual phenomena. Here are some roots of mangroves growing above the ground against gravitation, inhaling air with their fellow stem. Common passers-by or sailors, whosoever spots them, never know which they are, but get attracted with their staring view towards the sky with a deep sense of concentration. They are breathing roots, *Pneumatophore,* which becomes active inhaling oxygen to continue physiology of the body for survival. Seeds germinate in soil, grow, enlarge, but who knows seeds to grow while being attached with mother plant. Mangroves have the answer. *Vivipary* – a method of giving birth – mangroves adapt to nurture their progeny into their safe custody unless they are ready to fight against hungry tides and impairing salinity. In such beautiful forests, birds roost the canopy to twitter, monkeys to eat fruits, deer to graze on ground, crabs to hide into hole and many others to stay here to make a complete sphere of an abode at the edge of the land. However, it is a question of interest why they are so. Are they pushed away by their fellow land communities to be here or do they like to be here to behold the beauty of sea? Or they thought much before that the latest species – human beings – would continue their torture upon them – a speechless life. So they thought of growing near to the sea, escaping human savagery. Probably, nobody knows why they are here.

1.2   DEFINITION OF MANGROVES

The term 'Mangrove' usually defines the vegetation that occurs in the intertidal zones of coastal estuary in the tropical regions of the world (Figure 1.1). Mangroves comprise taxonomically diverse plant communities, including tree, shrub and palm species. The Oxford dictionary mentioned the word, 'mangrove' since 1613, referring to the tropical trees and shrubs growing in coastal swamps with tangled roots which grow above ground. Mangroves are salt-tolerant plant species growing at the land–sea interface, which connects interdependent links between inland terrestrial landscape and near shore marine environment. Experts opine that an individual plant of tidal forest refers to 'mangrove' whereas tidal plant communities along with the entire ecosystem are considered as 'mangal'. There are number of salt-tolerant species known as 'halophytes'. These species are herbaceous, grow across mangroves habitats from inner to outer estuary and exhibit prevalent vegetation.

FIGURE 1.1   A landscape view of mangrove forest in the Sundarban.

1.3   AN OVERVIEW

This book is prepared as a field guide and so avoids discussing the critical analyzes of terminologies which vary among scientists. However, the different terminologies applied to mangroves by the scientists need to be mentioned which readers be acquainted with.

1.3.1   GENESIS OF THE TERMS: MANGROVE AND MANGAL

No record is available when the word 'mangrove' came first in existence and who coined it. MacNae (1968) narrated a note on the genesis of mangroves, which is cited here to let workers know what mangrove is. It consists of compound words: the Portuguese word, 'mangue' and the English word 'grove.' The word mangue usually refers to an individual tree or shrub. The French 'manglier' is also used to apply individual kinds of tree. However, the word 'mangal' is commonly used to consider the forest community of trees, while the word 'mangroves' are referring to individuals' kinds of tree.

Keeping aside the European story, vernacular euphony of mangroves in the Asian context is also available. MacNae (1968) conceptualized that all these words seem to be akin to the word *manggi-manggi*, not existed now, to mangroves in Malay. In eastern Indonesia, the word as *Ambon* is used for *Avicennia* spp. The locals used to name *bada-ban* to mangroves in 'Sundarban'—the world largest single mangrove block situated in Ganga-Brahmaputra-Meghna estuary. Still the euphony *bada-ban* is popular among common people in the Indian part of the Sundarban. The word *bada* might be evolved from the distortion of the local dialect *kada* meaning mud and *ban* refers to the forest – the mud forest.

1.3.2   EARLY RECORDS OF MANGROVES – FOSSIL EVIDENCE

A general agreement suggests that mangrove ecosystems first appeared in the late Cretaceous-Early Tertiary on the shores of Tethys Sea (Ellison et al., 1999). Modern genera of mangroves arose on the eastern shores of Tethys, diversified towards present-day IWP (Indo-West Pacific) regions and then dispersed into AEP (Atlantic East Pacific) regions 3 million years ago (Steenis, 1962). The 'Center-of-origin hypothesis' suggests that conditions for the invasion of mangroves habitat occurred primarily in southeast Asia/Malaysia throughout most of the Tertiary, restricting most mangrove taxa to the IWP because of poor dispersal ability. During the mid-Tertiary mangroves dispersed to AEP because of the closure of Tethys connection to the Atlantic (Ricklefs and Latham, 1993). They further suggested that southeast Asia/Malaysia might act as more of a refugium than a center of origin. However, this hypothesis differs among scientists.

Scientists have a general view that taxonomically diverse species exhibit a common pattern of convergent adaptations to saline and anoxic habitats. These adaptations include aerial roots system, succulent sclerophyllous leaves and viviparous seedlings, most suited traits in sustenance of saline conditions (Ellison et al., 1999). Therefore, mangroves communities are much more an ecological assemblage rather than a taxonomic or morphological grouping (Saenger, 2002).

1.3.3 EARLY RECORDS OF MANGROVES – HISTORICAL EVIDENCES

Mangroves have been studied since ancient times. *Rhizophora* trees, true mangroves, found in the Red Sea and in the Persian Gulf were the earliest known records as described by Nearchus (325 B.C.) and Theophrastus (305 B.C). Plutarch (70 A.D.) and Abou'l Abass (1230) also recorded about *Rhizophora* and its seedlings (MacNae, 1968; Chapman, 1976). Rollet (1981) made the bibliography of mangrove research wherein he mentioned only 14 references before 1600, 25 references from the seventeenth century, 48 references in the eighteenth century, and 427 in the nineteenth century (Kathiresan and Bingham, 2001). On the contrary, there were 4500 mangrove references between 1900 and 1975 and approximately 3000 between 1978 and 1997, illustrating the explosion of interest in mangroves (Kathiresan and Bingham, 2001).

Vannucci (1997) also mentioned that the British learned the knowledge on how to manage mangroves for commercial timber production at Sunderban by the locals in the nineteenth century. The Portuguese, probably, the first Europeans to visit the mangrove forests of the Indian Ocean during the fourteenth century and learned the technique of rice-fish-mangrove farming. This Indian technology was also transferred to the African countries of Angola and Mozambique by Jesuit and Franciscan Fathers some six centuries ago (Vannucci, 1997).

1.3.4   EARLY RECORDS OF MANGROVES – MYTHOLOGICAL EVIDENCES AND BELIEF

Vannucci (1997) mentioned that mangroves have a long traditional link with human culture. In the Solomon Islands, the dead bodies are disposed of and special rites are performed in the mangrove waters. During the third century, in the peninsular region of India, a Hindu temple to the mangrove *Excoecaria agallocha* was erected (Kathiresan and Bingham, 2001). Mangroves plants being worshipped anciently as a 'sacred grove' are found in rock carvings and even today it is believed that a dip in the holy pond of the temple cures leprosy. Evidently, the city where this temple is found standing bears the name of the mangrove. In Kenya, the local people worship the shrines which were built in the mangrove forests. They believe that spirits of the shrine will bring death to those who cut the surrounding trees (Kathiresan and Bingham, 2001).

1.3.5   TERMINOLOGY OF MANGROVES AND THEIR CATEGORIES

The different plant communities constitute mangroves. Also, there are categories in mangroves. The criteria by which they are categorized into distinct groups vary among the scientists. Mangroves that constitute inter-tidal forest are prefixed with the words: 'major' or 'true' or 'compulsory' or 'obligate' or 'exclusive'. These prefixes are used by scientists to explain individual plants growing in intertidal zones. These plant communities are characterized with attributes which are adapted and befitted in saline habitats. Also, mangroves vary in numbers as assessed by different scientists. Tomlinson (1986) considered a total of 48 mangroves existing in the world and coined them to 'major mangroves' which are referred to 'true mangroves' or 'compulsory mangroves' or 'obligate mangroves' or 'exclusive mangroves' by others. There is a question of interest about which plant communities belong to the above groups. Saenger (2002) opined that true mangroves require freshwater for physiological activity while salt water is an ecological requirement. The former prevents excess respiratory losses and the latter prevents invasion and competition from other communities. Critical discussion and arguments are made since long by scientists. Naskar and Mandal (1999) studied morpho-anatomical characters of plant communities of intertidal forest and identified selected attributes adapted to those species growing exclusively in high salinity, with tidal fluctuations. They considered these species as 'Major mangroves' as mentioned in Table 1.1 (Mandal and Naskar, 2008). Wang et al. (2011) advocated that 'obligate mangroves' grow in saline conditions because of their physiological requirement. They highlighted that 'obligate mangroves' while grown in non-saline condition exhibited poor growth compared to saline habitats.

On the other hand, the woody plant communities, growing adjacent areas of intertidal zones, are coined as 'mangrove associates' or 'minor mangroves' or 'facultative mangroves' or 'semi mangroves' by scientists (Table 1.1). These plant communities lack most of the attributes, which are required for them to sustain in high salinity. Their few characters partially resemble the attributes of true mangroves, but are not conspicuous.

TABLE 1.1   The Intertidal Flora of the Indian Habitats and Their Category, Family and Lifeform

Category	Sl. No.	Scientific name	Family	Lifeform
Major mangroves	1	*Acanthus ebracteatus* Vahl	Acanthaceae	S
	2	*Acanthus ilicifolius* L.	Acanthaceae	S
	3	*Aegialitis rotundifolia* Roxb.	Agialitidaceae	S
	4	*Aegiceras corniculatum* (L.) Blanco	Myrsinaceae	S
	5	*Avicennia alba* Blume	Avicenniaceae	T
	6	*Avicennia marina* (Forssk.) Vierh.	Avicenniaceae	T
	7	*Avicennia officinalis* L.	Avicenniaceae	T
	8	*Bruguiera cylindrica* (L.) Blume	Rhizophoraceae	T
	9	*Bruguiera gymnorhiza* (L.) Savigny	Rhizophoraceae	T
	10	*Bruguiera parviflora* (Roxb.) Wight & Arn. ex Griff.	Rhizophoraceae	T
	11	*Bruguiera sexangula* (Lour.) Poir.	Rhizophoraceae	T
	12	*Ceriops decandra* (Griff.) Ding Hou	Rhizophoraceae	S
	13	*Ceriops tagal* (Perr.) C.B. Rob.	Rhizophoraceae	S/T
	14	*Excoecaria agallocha* L.	Euphorbiaceae	T
	15	*Heritiera fomes* Buch.-Ham.	Sterculiaceae	T
	16	*Heritiera littoralis* Aiton	Sterculiaceae	T
	17	*Kandelia candel* (L.) Druce	Rhizophoraceae	S/T

TABLE 1.1   *(Continued)*

Category	Sl. No.	Scientific name	Family	Lifeform
	18	*Lumnitzera littorea* (Jack) Voigt	Combretaceae	T
	19	*Lumnitzera racemosa* Willd.	Combretaceae	S
	20	*Nypa fruticans* Wurmb	Arecaceae	P
	21	*Phoenix paludosa* Roxb.	Arecaceae	P
	22	*Rhizophora apiculata* Blume	Rhizophoraceae	T
	23	*Rhizophora mucronata* Lam.	Rhizophoraceae	T
	24	*Rhizophora stylosa* Griff.	Rhizophoraceae	S/T
	25	*Scyphiphora hydrophylacea* C.F. Gaertn.	Rubiaceae	S
	26	*Sonneratia alba* Sm.	Sonneratiaceae	T
	27	*Sonneratia apetala* Buch.-Ham.	Sonneratiaceae	T
	28	*Sonneratia caseolaris* (L.) Engl.	Sonneratiaceae	T
	29	*Sonneratia griffithii* Kurz	Sonneratiaceae	T
	30	*Sonneratia ovata* Backer	Sonneratiaceae	T
	31	*Xylocarpus granatum* J. Koenig	Meliaceae	T
	32	*Xylocarpus moluccensis* (Lam.) M. Roem.	Meliaceae	T
Mangrove associates	1	*Acrostichum aureum* L.	Pteridaceae	F
	2	*Acrostichum speciosum* Willd.	Pteridaceae	F
	3	*Aglaia cucullata* (Roxb.) Pellegr.	Meliaceae	T
	4	*Ardisia elliptica* Thunb.	Myrsinaceae	S
	5	*Barringtonia racemosa* (L.) Spreng.	Lecythedaceae	T
	6	*Brownlowia tersa* (L.) Kosterm.	Tiliaceae	S
	7	*Carallia brachiata* (Lour.) Merr.	Rhizophoraceae	S
	8	*Cerbera odollam* Gaertn.	Apocynaceae	T
	9	*Clerodendrum inerme* (L.) Gaertn.	Verbenaceae	S
	10	*Cynometra iripa* Kostel.	Leguminosae	T
	11	*Dolichandrone spathacea* (L.f.) Seem.	Bignoniaceae	T
	12	*Hibiscus tiliaceus* L.	Malvaceae	T
	13	*Pemphis acidula* J.R. Forst. & G. Forst.	Lythraceae	T
	14	*Thespesia populnea* (L.) Sol. ex Corrêa	Malvaceae	T
Mangrove halophytes	1	*Acanthus volubilis* Wall.	Acanthaceae	Tw
	2	*Aeluropus lagopoides* (L.) Thwaites	Poaceae	G
	3	*Crinum defixum* Ker Gawl.	Amaryllidaceae	H

TABLE 1.1 *(Continued)*

Category	Sl. No.	Scientific name	Family	Lifeform
	4	*Cryptocoryne ciliata* (Roxb.) Fisch. ex Wydler	Araceae	H
	5	*Halosarcia indica* (Willd.) Paul G. Wilson	Amaranthaceae	H
	6	*Heliotropium curassavicum* L.	Boraginaceae	H
	7	*Hoya parasitica* Wall. ex Traill	Asclepiadaceae	Tw
	8	*Hydrophylax maritima* L.f.	Rubiaceae	Cr
	9	*Ipomoea pes-caprae* (L.) R. Br.	Convolvulaceae	Cr
	10	*Myriostachya wightiana* (Nees ex Steud.) Hook.f.	Poaceae	G
	11	*Pentatropis capensis* (L. f.) Bullock	Asclepiadaceae	Tw
	12	*Porteresia coarctata* (Roxb.) Tateoka	Poaceae	G
	13	*Salvadora persica* L.	Salvadoraceae	H
	14	*Sarcolobus carinatus* Griff.	Asclepiadaceae	Tw
	15	*Sarcolobus globosus* Wall.	Asclepiadaceae	Tw
	16	*Sesuvium portulacastrum* (L.) L.	Aizoiaceae	H
	17	*Suaeda maritima* (L.) Dumort.	Chenopodiaceae	H
	18	*Urochondra setulosa* (Trin.) C.E. Hubb.	Poaceae	G
Back mangals	1	*Caesalpinia bonduc* (L.) Roxb.	Leguminosae	Cl
	2	*Caesalpinia crista* L.	Leguminosae	Cl
	3	*Cynometra ramiflora* L.	Leguminosae	T
	4	*Dalbergia spinosa* Roxb.	Leguminosae	S
	5	*Dendrophthoe falcata* (L.f.) Ettingsh.	Loranthaceae,	E
	6	*Derris scandens* (Roxb.) Benth.	Leguminosae	Tw
	7	*Derris trifoliata* Lour.	Leguminosae	Tw
	8	*Flagellaria indica* L.	Flagellariaceae	Cl
	9	*Merope angulata* Swingle	Rutaceae	S
	10	*Pandanus furcatus* Roxb.	Pandanaceae	T
	12	*Pongamia pinnata* (L.) Pierre	Leguminosae	T
	12	*Solanum trilobatum* L.	Solanaceae	Cl
	13	*Tamarix gallica* L.	Tamaricaceae	S
	14	*Viscum orientale* Willd.	Loranthaceae	E

T, Tree; S, Shrub; H, Herb; G, Grass, E, Epiphyte; F, Fern; P, Palm; Cr, Creeper; Tw, Twiner, Cl, Climber.

In such discussion, we prefer to use the term 'mangrove halophytes' to the herbs and grasses which are salt-tolerant, grow exclusively in mangroves regions and get inundated most of the days a month (Table 1.1). They are prevalent in mangroves areas, and exhibit distinct morpho-anatomical features, including succulence, xeromorphy, and presence of sclerides, particularly in leaves adapted to withstand saline habitats. Another term 'back mangal' is widely known. These plant communities grow adjacent/ periphery to intertidal zones and get inundated only due to extremely tidal rise (Table 1.1). Their morpho-anatomical attributes are similar to those of mesophytes.

1.4   FEATURES OF MANGROVES

We mention a few distinctive features of different categories of mangroves and their habitats (Table 1.2). Workers are expected to have preliminary idea about the 'major mangroves', primarily to identify them in field conditions, along with their habitats. Also, they are likely to distinguish major mangroves from others such as 'mangrove associates', 'mangrove halophytes', and 'back mangals', each with features of respective habitats. All these attributes, but anatomical ones, are visible in field conditions, and careful field workers may be able to find them and distinguish them from one to another among different groups of mangroves.

1.5   KEY TO GENERA OF MAJOR MANGROVES

This field guide is intended as a preliminary aid to the identification of mangroves based on their visible morphological characters distinct in fields. In the following chapters mangroves are identified with diagnostic characters, even presented with illustrations. Here are mentioned few features of mangroves up to generic level to introduce field workers with the constituents of such unique coastal vegetation and therefore, their detailed features are intentionally avoided. Visible and distinct morphological features of leaves, roots and reproductive organs of respective genera are used in this key, which are of practical importance for field workers to distinguish from one genus to another. Not all the categories

TABLE 1.2 Grouping of Mangroves-Based on Features of Their Habits and Habitats

Group	Habitats	Features		Remark
		Morphology	Anatomy	
Major mangroves	I. Soil is mostly anaerobic known as 'physiologically dry soil' II. Land surface stretches from mudflat to consolidated-stable land III. Get inundated twice a day IV. Exhibit usually water-logged condition V. Salinity remains in a wide range of 08–30 ppt	I. Succulent leaves II. Upper surface shiny and flattened III. Thick hairs in lower surface IV. Lenticels numerous and Gall V. Pneumatophores – above ground roots arising against gravitation VI. Other above ground roots, Including knee roots, stilt roots, buttress and aerial roots VII. Vivipary or cryptovivipary germination	I. Thick cuticle II. Multilayered epidermis/ III. Hypodermis IV. Salt glands V. Sunken stomata VI. Aqueous tissue VII. Palisade mesophyll VIII. Dense sclerides/stone cells IX. Greater number of vessels	Mangroves, which are considered to be major/true ones, grow in such habitats and exhibit most of the features across different organs such as leaves, stem, roots and reproductive parts.
Mangrove associates	I. Soil is partly anaerobic II. Land surface usually stretches after consolidated-stable land III. Get inundated during MHWS IV. Salinity remains in a range of 03–07 ppt	I. Succulent leaves to some extent II. Sparse hairs in lower surface of some species III. Lenticels occasionally present	I. Dense sclerides/stone cells	Species are considered to be mangroves associate due to their partially affinities to true mangroves.

TABLE 1.2 (Continued)

Group	Features			Remark
	Habitats	Morphology	Anatomy	
Mangrove halophytes	I. Soil is both aerobic and anaerobic II. Land surface stretches from mudflat to consolidated-stable land III. Get inundated twice a day IV. Salinity remains variable	I. Succulent leaves II. Xeromorphic leaves III. Presence of sclerides in leaves	I. Thick cuticle II. Hypodermis multicelled III. Sunken stomata III. Aqueous tissue IV. Palisade mesophyll V. Sclerides/stone cells	Herbs and grasses growing in such habitats and exhibiting distinct succulence, xeromorphy and presence of sclerides in leaves, are considered to be halophytes.
Back mangals	I. Get rarely inundated during EHWS	I. No distinct morphological features found in relation to saline habitats	I. No distinct anatomical features found in relation to saline habitats	Species are found growing in mangroves adjacent areas

MHWS = Mean high water spring; EHWS = extremely high water spring.

of mangroves are mentioned; only species of those genera recognized, by almost all the mangroves workers in the world, as true constituents of mangroves vegetation are cited.

1A.	Leaves compound	2
1B.	Leaves simple	4
2A.	With 2–3 pairs of leaf-lets	*Xylocarpus*
2B.	With several long leaf-lets arranged along stout long mid-vain	3
3A.	Stem rhizomatous, leaf bases submerged	*Nypa*
3B.	Stem cylindrical, basal leaflets modified thorns	*Phoenix*
4A.	Leaves opposite	5
4B.	Leaves alternate	12
5A.	Viviparous hypocotyl present	6
5B.	Viviparous hypocotyl absent	9
6A.	Length of hypocotyl in the range of 20–75 cm	7
6B.	Length of hypocotyl in the range of 8–25 cm	8
7A.	Surface of hypocotyl with or without warty surface; broad lamina	*Rhizophora*
7B.	Surface of hypocotyl smooth, with pointed end; lanceolate lamina	*Kandelia*
8A.	Hypocotyl usually hanging from medium to large tree	*Bruguiera*
8B.	Hypocotyl usually hanging from small tree or developing upright from branch	*Ceriops*
9A.	Pneumatophore absent	10
9B.	Pneumatophore present	11
10A.	Texture of leaves coriaceous, margin with prickles	*Acanthus*
10B.	Texture of leaves succulent, with interpetiolar stipules	*Scyphiphora*
11A.	Pneumatophores abundant, pencil-like, coriaceous leaves	*Avicennia*
11B.	Pneumatophores rod-shaped, succulent leaves	*Sonneratia*
12A.	Leaves texture succulent, wavy margin	*Lumnitzera*
12B.	Leaves texture coriaceous	13
13A.	Milky latex present	*Excoecaria*
13B.	Milky latex absent	14
14A.	Stem without buttress, lamina lanceolate; white flower with strong aroma	*Aegiceras*
14B.	Stem with root buttress or swollen base	15
15A.	Leaves with long encircling petiole, stem with swollen base	*Aegialitis*
15B	Leaves texture leathery; distinct root buttress, plank-like extension	*Heritiera*

The presented key is likely to be easy for those who are already acquainted with botanical characters or at least some features of mangroves. It may be a bit of hard for beginners to follow, if they are not familiar to the unique features of mangroves. However, with the following chapters (i.e., Chapters 3, 4, and 5) readers are believed to be comfortable to know the mangroves with ease, when all the detailed, but concise, features of mangroves are unfolded like a bud to bloom with radiating beauty.

KEYWORDS

- **definition**
- **early records**
- **features**
- **terminology**

CHAPTER 2

DISTRIBUTION AND DIVERSITY OF MANGROVES

2.1 DISTRIBUTION OF MANGROVES IN THE WORLD

Two bio-geographic regions are broadly considered for distribution of mangroves worldwide (Figure 2.1), the main centers of diversity (Tomlinson, 1986): Indo-West Pacific (IWP) and Atlantic-East Pacific (AEP). Indo-West Pacific – the Eastern hemisphere known as The Old World tropic, is situated in between the longitudes 30°0′E–170°0′E, and spreads over three regions such as Australasia, Indo-Malesia, and Eastern Africa. Atlantic-East Pacific – the Western hemisphere is known as The New World tropic lying within the longitudes 15°0′E–120°0′W, which comprises also three regions such as Western Africa, Eastern America, and Western America. These two bio-geographic regions exhibit distinct diversification of mangroves. However, six mangroves bio-geographic regions (Table 2.1), situated in two hemispheres (IWP and AEP), are commonly considered for diversification of mangroves by experts (Saenger et al., 1983; Ricklefs and Latham, 1993; Spalding et al., 1997; Duke et al., 1998; Saenger, 2002).

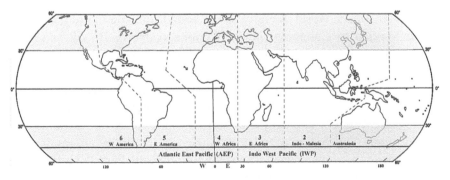

FIGURE 2.1   Map shows six bio-geographic regions of two hemispheres in the world.

TABLE 2.1   Latitudes Boundary of Mangroves on Major Land Masses

Hemisphere	Bio-geographic region	Northern limit (N)	Southern limit (S)
IWP	1. Australasia	—	38°45′
	2. Indo-Malesia (Pacific Asia)	31°22′	—
	3. Eastern Africa & Madagascar	27°40′	32°59′
AEP	4. Western Africa (eastern Atlantic)	19°50′	12°20′
	5. Eastern America (western Atlantic)	32°20′	28°56′
	6. Western America (Pacific America)	30°15′	5°32′

Source: Spalding et al., (1997, p. 29); Used with permission from ISME.

The total area of mangroves in the world was 181,077 km² as estimated by Spalding et al. (1997). FAO (2007) estimated about 154,704.5 km² mangroves area in the world, about 14.5% less (approx.) from the previous study. The latest study estimated the total mangroves forest area of the world as 137,760 km² covering 118 countries, 12% less (approx.) from the FAO estimation, which accounts for 0.7% of total tropical forests of the world; approximately 75% of mangroves are located in the 15 countries (Table 2.2) (Giri et al., 2011).

2.2   FEATURES OF MANGROVES HABITATS IN THE WORLD

Distribution of mangroves is restricted generally to the areas where mean air temperatures of the coldest month are higher than 20°C and where the seasonal range does not exceed 10°C (Walsh, 1974; Chapman, 1977). Mangroves habitats are classified into four groups: (i) Warm humid areas, where 90% of the world mangroves are found, include the regions from South Mexico to Colombia, in the Caribbean, North Brazil, and from South East Asia to north Queensland (Australia); (ii) Sub-humid areas, where mangroves are occasionally found, include East Africa, India, south Queensland (Australia), Mexico, and Venezuela; (iii) Semi-arid areas, where mangroves are rarely found, include Indus delta (Pakistan), Gujarat (India), the Western and northern Territory provinces of Australia, and Ecuador; and (iv) Arid areas, where mangroves are almost absent, include Ethiopian and Egyptian coastlines of the Red Sea, the Persian Gulf and the Gulf of California (Blasco, 1984).

TABLE 2.2 Biogeographic Regions Exhibiting Area (%) Along With Area Coverage (ha) of 15 Countries in the World

Hemisphere	Continent	Total area (%)		Country[*]	Area coverage[*]	
		Spalding et al. (1997)	Giri et al. (2010)		Area (ha)	Global total (%)
The Eastern (IWP)	Asia	42	42	Indonesia	3,112,989	22.6
				Malaysia	505,386	3.7
				Myanmar	494,584	3.6
				Bangladesh	436,570	3.2
				India	368,276	2.7
				Philippines	263,137	1.9
	Oceania	10	12	Australia	977,975	7.1
				Papua New Guinea	480,121	3.5
	Africa (East)	05	04[**]	Madagascar	278,078	2.0
				Mozambique	318,851	2.3
The Western (AEP)	Africa (West)	16	16[**]	Guinea Bissau	338,652	2.5
				Nigeria	653,669	4.7
	America (East and West)	27	26	Brazil	962,683	7.0
				Mexico	741,917	5.4
				Cuba	421,538	3.1

(Reprinted/modified with permission from Giri, C.; Ochieng, E.; Tieszen, L. L.; Zhu, Z.; Singh, A.; Loveland, T.; Masek, J.; Duke, N. Status and distribution of mangrove forests of the world using earth observation satellite data. Global Ecol. Biogeogr. 2011. © 2011 John Wiley.);

[**] African mangroves cover about 20% areas (East Africa, 4% + West Africa, 16%).

Mangroves vegetation is found luxuriant in the areas governed with certain environmental features:

- The annual rainfall ranges between 1500–2500 mm.
- Riverine input of freshwater discharge is found substantial.
- Moderate temperature ranges between 20–35°C.

On the contrary, aridity in the mangroves habitats is a limiting factor, resulting in stunted growth and sparse vegetation of the mangroves trees. The degree of aridity is defined as the basis on the values of ratio P/Etp, where P is the mean annual rainfall and Etp is the mean annual potential evapotranspiration. The extreme aridity of the regions is considered when the value of P/Etp is < 0.03 (Blasco, 1984).

2.3 MANGROVES DIVERSITY IN THE WORLD

The total number of intertidal plant communities as mangroves varies among scientists; because the criteria based on which the term 'mangrove' is considered differ among scientists. Chronologically here are six important assessments during the last 30 years on the total number of mangroves distributed worldwide.

1. Tomlinson (1986): A total number of true mangroves species are 48 found worldwide, of which 40 species represent from the IWP, only 8 species from the AEP, with distinct as well as respective mangroves species in the two world regions.
2. Ricklefs and Latham (1993): A total of 54 mangroves species are found worldwide. In IWP, Indo-Malesia shows 39 species, the highest numbers of mangroves diversity, followed by Australia & New Guinea with 35 mangroves and East Africa with 9 mangroves, being the least number. In AEP, West America shows 7 mangroves, followed by East America with 6 mangroves and West Africa with 5 mangroves, being the least mangroves diversity in both IWP and AEP.
3. Duke et al. (1998): A total of 70 mangroves species are found worldwide, of which 58 from IWP and 13 naturally occurring species from AEP, including one common species (*Acrostichum aureum*) found in both the hemispheres. In IWP, Indo-Malesia shows 51

species, the highest mangroves diversity, followed by Australia with 47 mangroves and East Africa with 11 mangroves, being the least. In AEP, West America shows 13 mangroves, followed by East America with 11 mangroves and West Africa with 8 mangroves, the least species diversity recorded from both the hemispheres.

4.   Saenger (2002): A total of 84 mangroves species are found worldwide, of which 68 from IWP and 19 naturally occurring species from AEP, including three common species (*Acrostichum aureum, Hibiscus tiliaceus* and *Thespesia populnea*) recorded from both the hemispheres. In IWP, Indo-Malesia shows 60 species, the highest mangroves diversity, followed by Australia with 54 mangroves and East Africa with 18 mangroves, being the least. In AEP, West America shows 18 mangroves, followed by East America with 16 mangroves and West Africa with 11 mangroves, the least species diversity recorded from both the hemispheres.

5.   Triest (2008): The Indo-Malaysian region of the IWP shows higher species diversity up to 40 compared to low species diversity less than 10 in the AEP. The greater species diversity in the IWP explains the more complex drift of tectonic fragments than that in the AEP. The species richness value was assessed as 41 (65%) in the IWP (south and southeast Asia and Australia), 8 (13%) in the Caribbean and eastern Pacific, 5 (8%) in West Africa, 9 (14%) in East Africa.

6.   Polidoro et al., (2010): A total of 70 species, considered as mangroves, are found worldwide; of which 36–46 species are occurring in the Indo-Malay-Philippines regions. The species diversity is greater in IWP than in AEP, suggesting that the distribution of mangroves richness declines from IWP to AEP.

2.3.1   ASSESSMENT OF MANGROVES DIVERSITY

Considering the number of mangroves listed by different workers, probably it is difficult to reach a general consensus on what should be the total number of mangroves worldwide. Mandal and Naskar (2008) classified mangroves of Indian habitats into three groups such as (i) Major mangroves, (ii) Mangrove associates and (iii) Back mangals, based on exclusively morpho-anatomical features. The study revealed the morphological as well as anatomical modifications of four organs (root, shoot, leaf and reproductive part) in relation to sustenance of plant communities growing in intertidal habitats. The species exhibiting greater degree of modifications of organs essential to survival strategy in intertidal habitats are considered to be major mangroves and those with less modified organs considered to be mangrove associates. Back mangals do not exhibit any modification of organs, but are mere communities growing in peripheral boundary of intertidal zones. Though the study of morpho-anatomical features was restricted only in Indian mangroves, such evaluation seems to be effective to classify plant communities growing in and around intertidal regions worldwide in consideration of mangroves.

We consider a total of 77 intertidal plant species as mangroves distributed in 37 genera and 25 families worldwide; of which 52 species belong to major mangroves and 25 species to mangrove associates (Table 2.3). A total of 65 species in 32 genera and 24 families are recorded from IWP, including three bio-geographic regions such as Australasia/Oceania, Indo-Malesia, and Eastern Africa. On the other hand, a total of 15 species in 10 genera and 8 families are recorded from AEP, including three bio-geographic regions such as Western Africa, Eastern America, and Western America. The naturally occurring species are included in this present list, but introduced species are avoided. Comparatively, IWP shows higher diversity than AEP. Among three bio-geographic regions of IWP, Indo-Malesia shows the highest diversity (58 species, 31 genera and 23 families), followed by Australasia/Oceania (51 species, 26 genera and 23 families) and Eastern Africa (16 species, 15 genera and 12 families) showing the poorest diversity. On the other hand, among three bio-geographic regions of AEP, Western America shows the highest diversity (14 species, 10 genera and 8 families), followed by Eastern America (13 species, 9 genera and 7 families) and Western Africa (9 species, 7 genera and 5 families) showing the poorest diversity.

With the evaluation of morpho-anatomical features of intertidal plant communities, these 77 species are reasonably categorized into two groups, e.g., major mangroves and mangrove associates (Table 2.3). Despite a number of important enumerations of mangroves by experts (Tomlinson, 1986; Duke et al., 1998; Naskar and Mandal, 1999; Saenger, 2002; Mandal and Naskar, 2008), we are encouraged to follow the web, www. theplantlists.org for updating mangroves list. In this web, few species which other workers mentioned as mangroves are either synonym or still unresolved to their status (discussed separately). When distribution of mangroves is dealt with worldwide, *Acrostichum aureum* is overlapping both the hemispheres (IWP and AEP). Morpho-anatomical study of *A. aureum* suggests that it be placed to mangrove associates, along with other two related species such as *A. speciosum* and *A. danaetfolium*. Both the latter ones are considered to be mangrove associates based on speculative morpho-anatomical features in supposedly relation to *A. aureum*. In a distributional array, two genera such as *Avicennia and Rhizophora,* major constituents of mangroves vegetation, are common in six biogeographic regions. Saenger (2002) listed an additional two species, *Hibisus tiliaceus* and *Thespesia populnea* as mangroves, overlapping both the hemispheres. The present study considers them mangrove associates. Further, a definite omission needs to be addressed that *Phoenix paludosa,* a mangrove palm, is missing from the list of the world mangroves species diversity (Saenger, 2002; Duke et al., 1998). This species is an important constituent of the Sundarban (Naskar and Mandal, 1999; Rahman et al., 2015), Bhitarkanica and Andaman & Nicobar Islands (Naskar and Mandal, 1999) mangroves vegetation sharing substantial mangroves areas in the Indo-Malesia regions. In persistent water-logged condition, *P. paludosa* develops above ground root (aerial roots), known as 'Pneumatothods', rising against gravitation. This palm is reasonably considered as 'true mangrove' with morpho-anatomical attributes exhibiting distinct modification to halophytic adaptation (Naskar and Mandal, 1999; Mandal and Naskar, 2008). We consider *Carallia brachiata* as mangrove associate for its morphological features and distribution coverage within intertidal regions; both criteria are supposed to exhibit close affinities to mangrove associates (Goutham-Bharathi et al., 2014). Two species of *Ceriops* such as *Ceriops zippeliana* Blume and *C. pseudodecandra* Sheue, Liu, Tsai & Yang (Rhizophoraceae) are not mentioned in the present list (Table 2.3), because only three species of *Ceriops* such as *C. australis, C. decandra* and *C. tagal* are widely known and accepted. However, Sheue et al. (2009, 2010) claims to have separated *C. decandra* into three species such as *C. decandra, C. pseudodecandra* and *C. zippeliana.* In this regard, the genus *Ceriops* comprises five species. The present list of intertidal plant communities (Table 2.3) includes major mangroves and mangrove associates separately after critical study of their morpho-anatomical attributes and distribution stretching over intertidal zones (Saenger, 1982, Saenger, et al., 1983; Tomlinson, 1986; Naskar and Mandal, 1999; Saenger, 2002; Mandal and Naskar, 2008). Nevertheless, we confess that the present list may suffer from errors and omissions.

TABLE 2.3 Distribution of World Mangroves, Lifeform, Red List Category and Global Loss

Category	Sl. no	Scientific name	Family	Lifeform	IWP			AEP			*Red List category	*Global loss (%)
					1	2	3	4	5	6		
Major mangroves	1	*Acanthus ebracteatus* Vahl	Acanthaceae	S	●	●					LC	22
	2	*Acanthus ilicifolius* L.	Acanthaceae	S	●	●					LC	20
	3	*Aegialitis annulata* R. Br.	Aegialitedaceae	S	●	●					LC	24
	4	*Aegialitis rotundifolia* Roxb.	Aegialitedaceae	S		●					NT	24
	5	*Aegiceras corniculatum* (L.) Blanco	Myrsinaceae	S	●	●					LC	21
	6	*Aegiceras floridum* Roem. & Schult.	Myrsinaceae	S		●					NT	29
	7	*Avicennia alba* Blume (synonym)	Avicenniaceae	T	●	●					LC	24
	8	*Avicennia bicolor* Standl.	Avicenniaceae	S/T						○	VU	31
	9	*Avicennia germinans* (L.) L.	Avicenniaceae	T					○	○	LC	17
	10	*Avicennia integra* N.C. Duke	Avicenniaceae	S/T	●						VU	<5
	11	*Avicennia marina* (Forssk.) Vierh.	Avicenniaceae	T		●	●				LC	21
	12	*Avicennia officinalis* L.	Avicenniaceae	T	●	●					LC	24
	13	*Avicennia rumphiana* Hallier f.	Avicenniaceae	T	●	●					VU	30
	14	*Avicennia schaueriana* Stapf & Leechm. ex Moldenke	Avicenniaceae	T					○		LC	6
	15	*Bruguiera cylindrica* (L.) Blume	Rhizophoraceae	T	●	●					LC	24
	16	*Bruguiera exaristata* Ding Hou	Rhizophoraceae	S/T	●						LC	23
	17	*Bruguiera gymnorhiza* (L.) Savigny	Rhizophoraceae	T	●	●	●				LC	20
	18	*Bruguiera hainesii* C.G. Rogers	Rhizophoraceae	T	●	●					CR	27
	19	*Bruguiera parviflora* (Roxb.) Wight & Arn. ex Griff.	Rhizophoraceae	T	●	●					LC	21
	20	*Bruguiera sexangula* (Lour.) Poir.	Rhizophoraceae	T	●	●					LC	21

TABLE 2.3 *(Continued)*

Category	Sl. no	Scientific name	Family	Lifeform	IWP 1	2	3	AEP 4	5	6	*Red List category	*Global loss (%)
	21	*Camptostemon philippinense* (S. Vidal) Becc.	Bombacaceae	T		●					EN	30
	22	*Camptostemon schultzii* Mast.	Bombacaceae	T	●						LC	24
	23	*Ceriops australis* (C.T. White) Ballment, T.J. Sm. & J.A. Stoddart	Rhizophoraceae	S/T	●						LC	24
	24	*Ceriops decandra* (Griff.) Ding Hou	Rhizophoraceae	S	●	●					NT	12
	25	*Ceriops tagal* (Perr.) C.B. Rob.	Rhizophoraceae	S/T	●	●	●				LC	18
	26	*Excoecaria agallocha* L.	Euphorbiaceae	T	●	●	●				LC	21
	27	*Heritiera fomes* Buch.-Ham.	Sterculiaceae	T		●					EN	50–80
	28	*Heritiera littoralis* Aiton	Sterculiaceae	T	●	●	●				LC	20
	29	*Kandelia candel* (L.) Druce	Rhizophoraceae	S/T		●					LC	23
	30	*Laguncularia racemosa* (L.) C.F. Gaertn.	Combretaceae	S/T				○	○	○	LC	17
	31	*Lumnitzera littorea* (Jack) Voigt	Combretaceae	T	●	●	●				LC	22
	32	*Lumnitzera racemosa* Willd.	Combretaceae	S/T	●	●	●				LC	19
	33	*Nypa fruticans* Wurmb	Arecaceae	P	●	●					LC	20
	34	*Phoenix paludosa* Roxb.	Arecaceae	P		●	●				NT	14
	35	*Rhizophora apiculata* Blume	Rhizophoraceae	T	●	●					LC	20
	36	*Rhizophora ×harrisonii* Leechm.	Rhizophoraceae	S/T				○	○	○		
	37	*Rhizophora mangle* L.	Rhizophoraceae	S/T				○	○	○	LC	17
	38	*Rhizophora mucronata* Lam.	Rhizophoraceae	T	●	●	●				LC	20
	39	*Rhizophora racemosa* G. Mey.	Rhizophoraceae	T				○	○	○	LC	15
	40	*Rhizophora samoensis* (Hochr.) Salvoza	Rhizophoraceae	T	●						LC	20

TABLE 2.3 *(Continued)*

Category	Sl. no	Scientific name	Family	Lifeform	IWP			AEP			*Red List category	*Global loss (%)
					1	2	3	4	5	6		
	41	*Rhizophora stylosa* Griff.	Rhizophoraceae	S/T	•	⋮	⋮	⋮	⋮	⋮	LC	20
	42	*Scyphiphora hydrophylacea* C.F. Gaertn.	Rubiaceae	S	•	•	⋮	⋮	⋮	⋮	LC	20
	43	*Sonneratia alba* Sm.	Sonneratiaceae	T	•	•	•	⋮	⋮	⋮	LC	20
	44	*Sonneratia apetala* Buch.-Ham.	Sonneratiaceae	T	⋮	•	⋮	⋮	⋮	⋮	LC	7
	45	*Sonneratia caseolaris* (L.) Engl.	Sonneratiaceae	T	•	•	⋮	⋮	⋮	⋮	LC	20
	46	*Sonneratia griffithii* Kurz	Sonneratiaceae	T	⋮	•	⋮	⋮	⋮	⋮	CR	80
	47	*Sonneratia* ×*gulngai* N.C. Duke & Jackes	Sonneratiaceae	T	•	⋮	⋮	⋮	⋮	⋮		
	48	*Sonneratia* ×*hainanensis* W.C. Ko, E.Y. Chen & W.Y. Chen	Sonneratiaceae	T	•	⋮	⋮	⋮	⋮	⋮		
	49	*Sonneratia lanceolata* Blume	Sonneratiaceae	T	•	•	⋮	⋮	⋮	⋮	LC	24
	50	*Sonneratia ovata* Backer	Sonneratiaceae	T	•	•	⋮	⋮	⋮	⋮	NT	28
	51	*Xylocarpus granatum* J. Koenig	Meliaceae	T	•	•	•	⋮	⋮	⋮	LC	21
	52	*Xylocarpus moluccensis* (Lam.) M. Roem.	Meliaceae	T	•	•	•	⋮	⋮	⋮	LC	21
Mangroves associates	1	*Acanthus volubilis* Wall.	Acanthaceae	S	•	•	⋮	⋮	⋮	⋮		24
	2	*Acanthus xiamenensis* R.T. Zhang	Acanthaceae	S	⋮	•	⋮	⋮	⋮	⋮		34
	3	*Acrostichum aureum* L.	Pteridaceae	F	•	•	•	○	○	○		19
	4	*Acrostichum danaeifolium* Langsd. & Fisch.	Pteridaceae	F	⋮	•	⋮	○	○	○		17
	5	*Acrostichum speciosum* Willd.	Pteridaceae	F	•	⋮	⋮	⋮	⋮	⋮		21
	6	*Aglaia cucullata* (Roxb.) Pellegr.	Meliaceae	T	⋮	•	⋮	⋮	⋮	⋮		23
	7	*Ardisia elliptica* Thunb.	Myrsinaceae	S	⋮	•	⋮	⋮	⋮	⋮		

TABLE 2.3 *(Continued)*

Category Sl. no	Scientific name	Family	Lifeform	IWP 1	IWP 2	IWP 3	AEP 4	AEP 5	AEP 6	*Red List category	*Global loss (%)
8	*Barringtonia racemosa* (L.) Spreng.	Lecythidaceae	T	●	●	●	⋮	⋮	⋮		
9	*Brownlowia tersa* (L.) Kosterm.	Tiliaceae	S	⋮	●	⋮	⋮	⋮	⋮		26
10	*Carallia brachiata* (Lour.) Merr.	Rhizophoraceae	S	⋮	●	⋮	⋮	⋮	⋮		
11	*Cerbera floribunda* K. Schum.	Apocynaceae	S	●	⋮	⋮	⋮	⋮	⋮		
12	*Cerbera manghas* L.	Apocynaceae	T	●	●	⋮	⋮	⋮	⋮		
13	*Cerbera odollam* Gaertn.	Apocynaceae	S/T	●	⋮	●	⋮	⋮	⋮		
14	*Clerodendrum inerme* (L.) Gaertn. Synonym	Verbenaceae	S	●	●	●	⋮	⋮	⋮		
15	*Conocarpus erectus* L.	Combretaceae	S/T	⋮	⋮	⋮	○	○	⋮		17
16	*Cynometra iripa* Kostel.	Fabaceae	T	●	⋮	⋮	⋮	⋮	⋮		21
17	*Diospyros vera* (Lour.) A. Chev.	Ebenaceae	T	●	⋮	●	⋮	⋮	⋮		24
18	*Dolichandrone spathacea* (L.f.) Seem.	Bignoniaceae	T	●	⋮	●	⋮	⋮	⋮		23
19	*Hibiscus tiliaceus* L.	Malvaceae	T	●	⋮	●	○	○	○		
20	*Mora oleifera* (Hemsl.) Ducke	Fabaceaea	T	⋮	⋮	⋮	○	○	⋮		26
21	*Osbornia octodonta* F. Muell.	Myrtaceae	S/T	●	⋮	●	⋮	⋮	⋮		23
22	*Pelliciera rhizophorae* Planch. & Triana	Pellicieraceae	T	⋮	⋮	⋮	○	○	○		27
23	*Pemphis acidula* J.R. Forst. & G. Forst.	Lythraceae	S	●	●	●	⋮	⋮	⋮		21
24	*Tabebuia palustris* Hemsl.	Bignoniaceae	S	⋮	⋮	⋮	⋮	⋮	○		33
25	*Thespesia populnea* (L.) Sol. ex Corrêa	Malvaceae	T	●	●	●	○	○	○		—

T, tree; S, shrub; P, palm; CR, critically endangered; EN, endangered; VU, Vulnerable; NT, near threatened; LC, least concern; DD, data deficient.

(Reprinted from Polirado BA, Carpenter KE, Collins L, Duke NC, Ellison AM, Ellison JC, et al. (2010) The Loss of Species: Mangrove Extinction Risk and Geographic Areas of Global Concern. PLoS ONE 5(4): e10095. https://doi.org/10.1371/journal.pone.0010095. Creative Commons Attribution License.)

2.3.2   RELATIVE DIVERSITY OF MANGROVES

In the assessment of mangroves diversity (Table 2.3) in two hemispheres and six bio-geographic regions, we follow the index developed by Mandal and Naskar (2008) for relative mangrove diversity (RMD) =100 × [(Fn+Gn+Sn)/N], where Fn, Gn and Sn are, respectively, numbers of families, genera and species of each hemisphere or biogeographic region, and N = 139 (sum of reported numbers of families, genera and species from two hemispheres, e.g., 25+37+77). Given respective values (numbers) in diversity index, IWP scores 87.05% and AEP with 23.97%. Among six biogeographic regions, Indo-Malesia scores 80.57%, followed by Austral-asia/Oceania with 71.94%, Eastern Africa with 30.93%, Western America 23.02%, Eastern America 20.86% and Western Africa 15.10%.

2.3.3   MANGROVES EXTINCTION RISK: THREATENED SPECIES AND GLOBAL LOSS

As assessed (Polidoro et al., 2010), a total of 65 mangroves, out of 77 species, are listed in Red List of Threatened species, falling into 6 categories (Table 2.3). The status of other 12 species is not known. A total of 47 species are assessed to be the least concern (LC), followed by 6 species each to the vulnerable (VU) and the near threatened (NT), 2 species each to the data deficient (DD), the endangered (EN) and critically endangered (CR). Species falling to LC exhibit the global loss ranging between 6–24%. Two species are assessed as CR, with *Sonneratia griffithii* to 80% loss and *Bruguiera hainesii* to 27% loss. Other two species which are also in alarming state include *Heritiera fomes* with 50–80% and *Camptostemon philippinense* with 30% loss.

2.4   MANGROVES IN INDIA

India has a prominent place in the area coverage of mangroves vegetation in the world (Table 2.2), with long coastlines about 7516.6 km, including the island territories. The East Coast line (2700 km long) is situated along the Bay of Bengal, the West Coast line (3000 km long) along Arabian Sea and the Andaman and Nicobar Islands (a total of 1816.6 km) surrounded by Bay of Bengal. The mangroves of India are distributed in the geographical locations: longitudes between 69°E –89°5′E and latitudes between 7°N –23°N, with three distinct zones: (i) East Coast habitats, (ii) West Coast habitats, and (iii) Andaman and Nicobar Islands habitats. Three regions are distinct with the features of geomorphology following Thom's (1982) classification of coastal habitats (Mandal and Naskar, 2008). The East Coast belongs to deltaic mangroves habitats, the West Coast to coastal mangroves habitats, and the Andaman and Nicobar Islands to Island mangroves habitats.

India has 10 distinct mangroves habitats (Table 2.4). The following five major mangroves habitats are situated along the East Coast of India and in Bay of Bengal (Figure 2.2a):

1. Sundarban mangroves forest,
2. Bhitarkanica mangroves forest,
3. Godavari and Krishna delta mangroves forest,
4. Cauveri delta mangroves forest, and
5. Andaman and Nicobar Islands

The other five minor mangroves habitats situated along the West Coast of India (Figure 2.2b) are:

6. Cochin estuary,
7. Coondapoor estuary,
8. Zuary estuary,
9.   Bombay coast, and
10. Bhabnagar estuary.

2.4.1   FEATURES OF MANGROVES HABITATS

The features of geomorphology, hydrology and climate that constitute together as 'macro environment' are distinct between the East Coast and the West Coast of India. Andaman and Nicobar islands as situated in the Bay of Bengal may be considered as a part in the scale of the east coastal macro environment. East Coast is characterized with freshwater carrying major rivers, strong tidal current, network of deltas and moderate rainfall, with the value of *P/Etp* to be > 0.75 (Table 2.4). About 85% India's mangroves coverage is found in East Coast and in Andaman and Nicobar islands, with dense and luxuriant vegetation (Blasco and Aizpuru, 2002). On the other

TABLE 2.4 Features of Mangroves Habitats in India (Govt. of India, 1987; Naskar and Mandal, 1999; Blasco, 2002; Selvan, 2003; Kumar et al., 2005; Mandal and Naskar, 2008; CAS, 2011)

Region	Habitat (km²)	Geomorphology		Hydrology			Climate	
		Setting	Dominant Features	Source of freshwater	Annual average discharge (m³/s)	Tidal rise (m) in range	Annual rainfall (mm)	Dry month
East Coast	Sundarbans (2000)	Tide dominated allochthonous type of mangroves	High tidal range with strong bidirectional current	Ganga	38,129	5–8.0	1500–2000	4
	Bhitarkanica (50)			Mahanadi	2119	2–6.0	1500–2000	4
	Godavari & Krishna (100)	River dominated allochthonous type of mangroves	Rapid deposition of terrigenous material; delta expand seaward	Godavari & Krishna	3505 & 2213	0.8–1.3	1200	6
	Pichvaram and Mathupet (15)			Cauvery	677	0.64–0.9	800–1300	6–8
Islands	Andaman & Nicobar islands	Carbonate platform on low energy coasts type of mangroves	Platform slowly accreting due to the accumulation of marl (calcareous) and peat	Rain water	—	1.5–1.9	2750–3080	3
West Coast	Cochin estuary	Fringe type of mangroves	Beaches characterized with sand dunes, cliffs and silt clay	Minor rivers and Rain water	—	0.6–0.9	2600	8
	Coondapur & Malpe estuary					0.9–1.5	2600	8
	Zuary estuary					1.0–1.9	2600–3000	8
	Bombay coast					1.86–4.42	2500	8
	Bhavnagar estuary	Fringe type of mangroves with scattered trees	Bedrock valley drowned by rising sea-level			1.5–2.2	470–900	8

hand, West Coast is characterized with narrow and steep in slope, lack of freshwater flowing major rivers and persistence of dry months, with the value of ratio *P/Etp* ranging between 0.5–0.75. Mangroves are sparse and less diverse.

2.4.2 MANGROVES DIVERSITY IN INDIAN HABITATS

In species diversity, a total of 78 species recorded (Table 1.1 and Figure 2.2a and b) are distributed in 56 genera and 37 families, which constitute estuarine vegetation in and around intertidal zones, of which a total of 32 species are considered as major mangroves, 14 species as mangrove associates, 18 species as mangrove halophytes and 14 species as back mangals.

The present classification has been made based on morpho-anatomical features. Major mangroves, by and large, exhibit modification of four organs (root, shoot, leaf and reproductive part) befitted for halophytic adaptation. Mangrove associates have attributes of at least one organ modified and their modification occurs in leaf which shows either thick cuticle layer like true mangroves or water storing mechanism or prevention of water loss through transpiration by developing gregarious hairs. The study also prefers the term 'mangrove associates' to these species with the rationale that they are not mangroves but are mere associates while growing adjacent to intertidal zones. A total of 18 species of herb and of grass exhibit their prevalent presence in intertidal zone inundated with frequent tidal water. They show succulence prevalent in both leaf and stem suitable of coping with high salinity in intertidal zones. We consider all these species as 'halophytes' due to their herbaceous habit and so they are coined as 'mangrove halophytes.' However, they form definite intertidal vegetation and have been major constituent of fringe vegetation. Back mangal species do not exhibit any attribute modified or relating to halophytic adaptation. These plants are found growing beyond intertidal zones and rarely grow in adjacent to intertidal areas.

Comparatively, species diversity of East Coast mangroves habitats and Andaman & Nicobar Islands is higher than that of West Coast mangroves habitats. A total of 76 species are distributed in 54 genera and 36 families, comprising major mangroves, mangrove associates, mangrove halophytes and back mangals. There exists a significant disparity in the number of mangrove species reported in a given region which could be attributed to

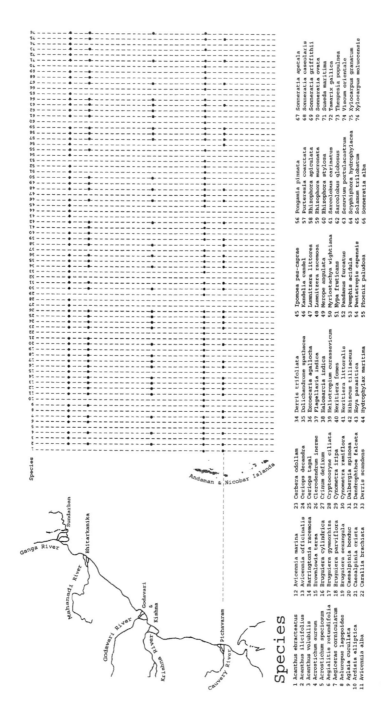

FIGURE 2.2a Distribution of mangroves in East Coastal habitats of India and Andaman & Nicobar Islands.

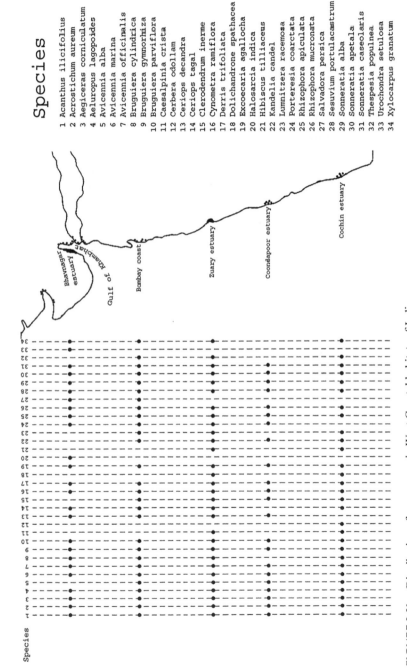

FIGURE 2.2b Distribution of mangroves in West Coastal habitats of India.

a number of unresolved taxonomic ambiguities. For instance, only two species of *Xylocarpus* (*X. granatum* and *X. moluccensis*) are regarded as true mangroves after critically reviewing the status of four species of the same genus previously reported from Andaman and Nicobar Islands (Goutham-Bharathi et al., 2014). Other two species such as *X. gangeticus* and *X. mekongensis* are the result of misidentification. *X. mekongensis,* which was the synonym of *X. moluccensis*, is confirmed from the web, www.theplantlists.org and *X. gangeticus*, probably, is a synonym of *X. rumphii* – a non mangrove species. Recently, Ragavan et al. (2015) reported two hybrids of *Rhizophora* such as *Rhizophora* × *annamalayana* Kathir and *Rhizophora* × *lamarckii* Montrouz from the Andaman and Nicobar Islands, India though former one was reported long before by Kathiresan (1995) and later supported by Lakshmi et al. (2002) from Pichavaram mangrove forest, Tamil Nadu, India. However, all those reports on new hybrids in India were based on only morphological differences, which seem to be ambiguous (Jayatissa et al., 2002); because taxonomical distinction among them was not much explored. Until recently, Sahu et al. (2015) claims that these two hybrids are considered to be new records, which has been established based on AFLP analysis. Sometimes, the ambiguities are surfaced regarding the enumeration of an actual number of mangroves because of lack of proper identification or may be the ecological variation of the same species while growing in different regions. To resolve the ambiguity, we suggest the latest technique like DNA fingerprinting to be the way out.

2.4.3   RELATIVE MANGROVES DIVERSITY OF INDIA

In the assessment of mangroves diversity (Table 1.1 and Figure 2.2a & b) of two regions, e.g., East Coast and Andaman and Nicobar Islands and West Coast, we like to modify our earlier report on the status of mangroves diversity (Mandal and Naskar, 2008) since numbers of species have been revised. In relative mangrove diversity (RMD) $=100 \times [(Fn+Gn+Sn)/N]$, where Fn, Gn and Sn are, respectively, numbers of families, genera and species of each region, and N = 171 (sum of reported numbers of families, genera and species from ten regions, e.g., 37+56+78). Given respective values (numbers) in diversity index, East Coast and Andaman and Nicobar Islands score 97.07% and West Coast with 45.61%. Among ten habitats of two regions, Sundarban mangroves forest scores 86.54%, Bhitarkanica mangroves forest 65.49%, Godavari and Krishna delta mangroves forest 46.19%, Cauveri delta mangroves forest 47.95%, Andaman and Nicobar Islands 80.11%, Cochin estuary 39.18%, Coondapoor estuary 28.65%, Zuary estuary 39.18%, Bombay coast 36.84% and Bhabnagar estuary 32.74%.

2.4.4   FEATURES OF MANGROVES VEGETATION

Mangroves vegetation is classified into 12 types based on their height and density (Blasco and Aizpuru, 2002). All these are not mentioned necessarily in this book; few distinctive types found in Indian habitats are discussed. Quite a few of mangroves type are prevalent in the Indian Sundarban, including 'tall dense mangroves forest' like *Avicennia* spp., 'low dense forest' like *Ceriops* spp., 'open mangroves thickets' like palm – *Phoenix paludosa*, and 'scattered mangroves under shrubs' like *Suaeda* sp. Importantly, Sundarban is the single largest mangroves block in the world (Giri et al., 2011), situated in the estuary of three major rivers, Ganga-Brahmaputra-Meghna. The almost similar features of mangroves vegetation appear in the Bhitarkanica forest, Mahanadi delta, but are not distinct as Sundarban. 'Tall open mangroves forest', 'open mangroves thicket' and 'low degraded mangroves' are found in Godavari & Krishna deltas. Cauvery delta is characterized with 'low open mangroves forest'. Comparatively, 'tall dense mangroves' with luxuriant growth are found in Andaman and Nicobar islands. Mostly degraded, scattered and low thickets mangroves are found in the habitats of West Coast of India due to prevalence of persistent aridity.

2.5   E-TAXONOMY IN NOMENCLATURE

The use of correct name for a taxon is the hallmark of nomenclature. The correct nomenclature of a taxon manifests its clarity in one way and mini-mizes ambiguity in other; because synonyms of a taxon develop confusion among users. It does not reflect the actual species identity as well as its taxonomical hierarchy, even composition of species remains uncertain in a given area. Particularly, biological researchers are greatly affected by such misidentification that may also lead to gross misinterpretation of respective biological entity. To resolve this problem to a great extent, there are quite a few resources now available in online database known as e-taxonomy, application of online database/ website used for validation of a name of a taxon according to Botanical Code. Some databases through which synonyms and accepted name of the taxa are searched are the result of a compilation of multi-checklist data with its significant marks. One can find the status of particular taxon, with its accepted name, synonyms and also know whether it is unresolved. Not that this online database/ website is entirely reliable, but that there has been updated information about a taxon through compilation of enormous data from various sources. Reader can verify the details of certain species by using such data web and may have an opportunity to establish reliable information for a particular species. Therefore, readers are suggested to search more online databases/ websites (http://www.catalogoflife.org and http://www.gbif.org) than one (www.theplantlist.org) to ensure the updated position of any taxon or its latest nomenclatural changes. For instance, in few mangrove species, the website, www.theplantlist.org fails to revise the following latest changes (see Chapter 5): *Ceriops decandra* (Griff.) Ding Hou has been the updated author citation instead of *Ceriops decandra* (Griff.) W. Theobald. Similarly, *Bruguiera gymnorhiza* (L.) Savigny has been the new author citation instead of *Bruguiera gymnorhiza* (L.) Lam. and *Hibiscus tiliaceus* L. has been wrongly spelled as *Hibiscus tilliaceus* L.

2.6   E-TAXONOMY IN NOMENCLATURE OF MANGROVES

Considering the importance of nomenclature, we feel reasonable to revise the mangroves list through e-taxonomy following the website, www.theplantlist.org – the working plant groups. The intention is to update the list of mangroves with new circumscription to any level of taxonomical hierarchy, particularly restricted to family, genus and species. Common and familiar names of taxa considered as old circumscriptions are mentioned, along with new circumscriptions (Table 2.5) that enable workers to understand the status of respective species to its updated position. For example, this guidebook has already used the new generic circumscription as *Halosarcia* instead of *Salicornia*. In the case of family the new circumscription is widely used in recent literatures as Avicenniaceae has been

TABLE 2.5 Revised Status of Taxa in Mangroves Following the Web: www.theplantlist.org

| Family | | Genus | | Species | |
Old circumscription	New circumscription	Old circumscription	New circumscription	Old circumscription	New circumscription/remarks
				Avicennia rumphiana Hallier f.	Synonym of *Avicennia marina* var. *rumphiana* (Hallier f.) Bakh.
				Rhizophora lamarckii Montrouz.	Unresolved
				Rhizophora selala (Salvoza) Toml.	Unresolved
				Sonneratia lanceolata Blume	Unresolved
				Sonneratia ovata Backer	Unresolved
				Solanum trilobatum L.	Unresolved
				Sarcolobus carinatus Griff.	Unresolved name
				Sarcolobus globosus Wall.	Unresolved name
				Crinum defixum Ker Gawl.	Synonym of *Crinum viviparum* (Lam.) R. Ansari & V.J. Nair
				Hoya parasitica Wall. ex Traill	Unresolved name
				Avicennia alba Blume	Synonym of *Avicennia marina* (Forssk.) Vierh.
Aegialitedaceae	Plumbaginaceae				
Amaryllidaceae					
Asclepiadaceae	Apocynaceae				
Avicenniaceae	Acanthaceae				
Bombaceae	Malvaceae				

TABLE 2.5 *(Continued)*

Family		Genus		Species	
Old circumscription	New circumscription	Old circumscription	New circumscription	Old circumscription	New circumscription/remarks
Chenopodiaceae	Amaranthaceae	*Salicornia*	*Halosarcia*	*Salicornia brachiata* Miq.	*Halosarcia indica* (Willd.) Paul G. Wilson
Loranthaceae	Santalaceae			*Viscum orientale* Willd.	*Viscum cruciatum* Sieber ex Boiss.
Myrsinaceae	Primulaceae				
Pellicieraceae	Tetrameristaceae				
Sonneratiaceae	Lythraceae			*Sonneratia griffithii* Kurz	Unresolved
Sterculiaceae	Malvaceae				
Tiliaceae	Malvaceae				
Verbenaceae	Lamiaceae	*Clerodendrum*	*Volkameria*	*Clerodendrum inerme* (L.) Gaertn.	Synonym of *Volkameria inermis* L.,

Unresolved names are those to which it is not yet possible to assign a Status of either 'Accepted' or 'Synonym.'

revised and replaced under Acanthaceae and Sonneratiaceae to Lythraceae and many more. This exercise may be helpful for workers and students to be aware of scientific revision of mangroves with recent information through e-taxonomy. More and above, field workers may be able to identify mangroves and mangrove associates with spirit of scientific quest whether they exist or disappear from their respective habitats, following this manual or other means.

KEYWORDS

- **distribution**
- **e-taxonomy**
- **features**

CHAPTER 3

ECOLOGY AND SUCCESSION OF MANGROVES

3.1 ECOLOGY

Ecology deals with the interaction between living and non-living components and also interaction among organisms in a given area. Since this book is about a field guide for identification of mangroves, we feel appropriate to avoid the details of mangroves ecology, leaving the necessary elements required during the field study. So this book is planned to mention only that part of ecology which needs to be like a guide for readers/workers to know about the possible position and location of mangroves in their habitats, and the factors responsible for such condition. The distance of tidal ingress, types of inundation, duration of inundation, and degree of water salinity – all these relevant environmental factors are discussed to simplify the occurrence of mangroves on land surface. Mangroves are distributed as much as regional specific, as governed by the intensity of the environmental factors. Therefore, habitats of mangroves are considered to be dynamic; for there are regular tidal rise and tidal flow to inundate land surface, resulting in degradation of land in one side and formation of mudflat in other – a view of eroded slope and extended flat surface. There is a question of why a pool of species existing in a particular location, and disappearing with time, replaced by other communities. There is also a question of interest to which factors determining the location and position of mangroves. Somewhere they are with dense population, and may be sparse in other locations. One may not find the same community of mangroves as permanent one that succeeded on a particular area over a couple of periods; because habitat conditions seem to be ever changing, might cause its sudden disappearance. The specific land surface exists today as a stable land, tomorrow which might be degraded one. This book attempts to mention all those relevant queries, trying to apparently satisfy readers' curiosities.

3.2 INTERTIDAL ZONE

Mangroves and intertidal zone are closely related in ecological perspective; major mangroves are restricted growing in intertidal zone (Figure 3.1a & b). The term 'Intertidal zone' refers to the area situated between two tidal limits, e.g., the 'high tide' and the 'low tide'. The occurrence of tidal rise and flow (the vertical difference between the high tide and the succeeding low tide), caused by the pull of the moon and the sun with the rotation of the earth, is known as 'Tidal fluctuation' or 'Tidal range'. Tidal rise occurs twice a day (24 h 52 min) regularly along the seacoast, estuary and deltaic Islands, with the time interval lying between two peak tides about 12 h 26 min a day (Figure 3.2a). The degree of tidal fluctuation is considered as what height a tide reaches, measured with different parameters based on the tidal limits: MSL (Mean Sea Level), MHWN (Mean High Water Neap), MHWS (Mean High Water Spring) and EHWS (Extra High Water Spring). Also, the strength of tidal fluctuation is determined as how far distance a tide travels over a land surface and inundates its vegetation. The height of the tidal rise and traverse of the tidal ingress to the distance from seashore, both are correlated.

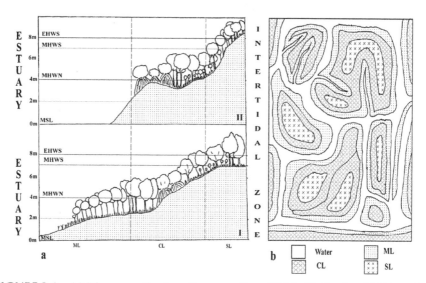

FIGURE 3.1 (a) Diagrammatic view of cross-section of I. Mudflat land and II. Degraded land of intertidal zone of mangroves forest; (b) The diagram shows a bird's eye view of the typical intertidal zone comprising ML (Mudflat land), CL (Consolidated land) and SL (Stable land) in the Indian Sundarban mangrove forest.

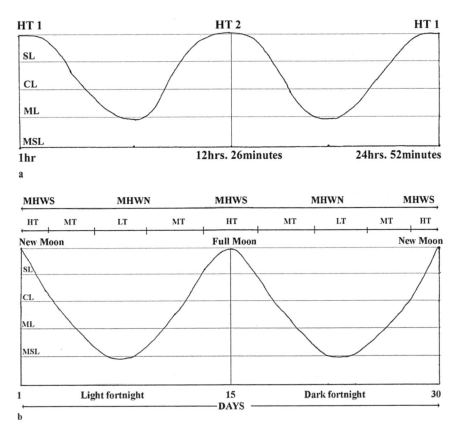

FIGURE 3.2 (a) The diagram shows tidal fluctuation of a day; (b) diagram shows tidal shifting along different days within fortnight (both light and dark fortnight).

3.3   ZONATION AND INUNDATION CLASS

Land surface, stretching from seaward to landward, exhibits distinct zonations. Readers need to know about the zonation, which is mostly regional set and location-specific of any intertidal zone. There are broadly three zones: Mudflat Land (ML), Consolidated Land (CL) and Stable Land (SL).

3.3.1   FEATURES OF LAND SURFACE

Mudflat land: It is found as newly formed soil strata, bordered with water edge at one side.

Consolidated land: It is considered as comparatively stiff land lying after mudflat zone.

Stable land: It is established land mass, situated after consolidated zone.

3.3.2 LAND SURFACE AND INUNDATION

Mudflat land, by and large, gets inundated in 28–31 days (56–62 times a month; since tidal rise occurs twice a day), Consolidated land in 14–21 days (28–42 times a month) and Stable land in 0–5 (0–10 times a month). Besides these categories, two transitional zones are considered to be lying in between (Table 3.1); one is Mudflat-consolidated land zone and another one as a consolidated-stable land zone. Mudflat-consolidated zone, lying between mudflat and consolidated land, gets inundated in 22–27 days (44–54 times/month) and Consolidated-stable land, lying between consolidated land and stable land, gets inundation in 6–13 days (12–26 times/month).

TABLE 3.1 Zonations Showing Land Pattern, in Relation to Tidal Types and Status of Inundation

Zone	Land pattern	Tidal type	Inundation/month	
			Days	Number
I	Mud flat land	All tides (LT, MT and HT)	28–31	56–62
II	Mud flat-consolidated land	MT and HT, LT (occasional)	22–27	44–54
III	Consolidated land	MT and HT	14–21	28–42
IV	Consolidated – Stable land	HT, MT (occasional)	6–13	12–26
V	Stable land	HT	0–5	0–10

Tidal inundation is the result of tidal ingress, indicating the distance of land surface the tidal rise travels. The time of tidal rise and flow shifts day wise on the particular place within fortnight (Figure 3.2b). EHWS travels extreme distance, though it is a rare occurrence. MHWS travels maximum distance towards land surface and occurs during near to pre and post 'Full moon' time or 'New moon' time, covering 2–3 days within a fortnight. Both EHWS and MHWS, known as 'High tide' (HT), inundate all 5 zones of land surface. After the peak of 'Full moon' or of 'New moon', tidal rise gradually loses its strength, indicating low rise and coverage of less distance comparatively. Considered to be 'Medium tide' (MT), such tidal rise is the effect of both MHWS and MHWN and inundates 3–4 zones. When tidal rise becomes weak and inundates 1–2 zones, it is known as 'Low tide' (LT), the effect of MHWN.

3.4   MANGROVES HABITATS ALONG ESTUARY

Tidal ingress occurs to reach its distance against water flow of river (Figure 3.3). The distance of river lying between the point of maximum limit of tidal ingress and the estuary mouth is categorized as (i) Upstream (US), (ii) Midstream (MS), and (iii) Downstream (DS). Mangroves are found growing in different positions of estuary, characterized with degree of salinity. Comparatively, low salinity (2–10 ppt) prevails in upstream and high salinity, gradually, dominates mid-stream (8–20 ppt) and then down-stream (18–30 ppt), being the highest saline zone. Different mangroves are adapted growing in specific regime of salinity occurring in habitats, based on the tolerance limit of respective species.

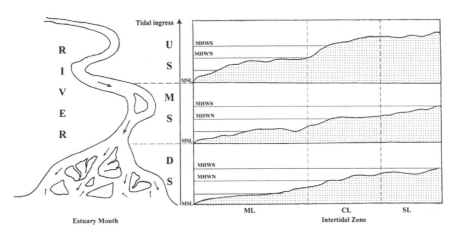

FIGURE 3.3   The diagram shows different positions of estuary such as US (upstream), MS (midstream) and DS (downstream), along with locations of respective intertidal zones such as, ML (Mudflat land), CL (Consolidated land) and SL (Stable land) in the Indian Sundarban mangrove forest.

3.5   STEPS TO KNOW THE POSITION OF ESTUARY

This may be difficult for anybody to know about what position of estuary among three categories the particular mangroves habitat belongs to. However, one should keep in mind few criteria: (i) distance of seashore, (ii) degree of water salinity, (iii) prominence of tidal ingress. In case one may discuss with the locals. We mention below some guidelines to identify the conditions of habitats along the position of estuary; these are not exclusive guidelines for consideration, rather some clues to proceed. Learners need to carry some minor equipments/materials while studying ecology in fields, which include the following:

   a)   Measuring tape for tidal ingression;
   b)   Folding measuring stick for tidal fluctuation;
   c)   Refractometer for salinity;
   d)   Small compass for geographical direction;
   e)   GPS for local positioning.

3.5.1   UPSTREAM ESTUARY

Step I: Observe the riverbank, if any mark of tidal rise is found.
Step II: See the vegetation, if stable land is dominated with mangrove associates and Back mangals.
Step III: Analyze water salinity ranging between 2–10 ppt, if possible.

3.5.2   MIDSTREAM ESTUARY

Step I: Observe tidal rise and flow to be distinct.
Step II: Dominance of major mangroves is found, with sparse mangroves associates.
Step III: Analyze water salinity ranging between 8–20 ppt, if possible.

3.5.3   DOWNSTREAM ESTUARY

Step I: Observe tidal rise and flow with sea surges.
Step II: Dominance of major mangroves.
Step III: Analyze water salinity ranging between 18–30 ppt, if possible.

3.6 STEPS TO KNOW THE ZONATION OF LAND SURFACE

Step I: Observe the land edge if it is newly silted soil strata adjoining sea, consider Zone −1 (mud flat), dominated with grass vegetation.

Step II: See the vegetation, which may usually comprise the following species: *Porteresia coarctata, Avicennia alba, A. marina,* consider it Zone II (mud flat -consolidated land). In some area *Sonneratia apetala* may be prominent.

Step III: Vegetation of *Avicennia* spp. may be sparse, rather major mangroves like *Rhizophora* spp., *Bruguiera* spp., *Ceriops* spp., *Sonneratia* are prominent among others, consider it Zone III (consolidated land).

Step IV: Vegetation is found to be mixed stand with major mangroves like *Rhizophora* spp., *Bruguiera* spp. *Aegiceras* sp., *Aegialitis* sp., *Heritiera* spp., *Xylocarpus* spp., *Sonneratia* prevalent among others, consider it Zone IV (consolidated-stable land).

Step V: May be palm trees are prominent among other major mangroves with mangroves associated species present in periphery, consider it Zone V (stable land).

The guidelines are mentioned as some steps on how to proceed for field study of mangroves, particularly for beginners. There are many criteria overlapping one another, and sometimes not matching properly. Nevertheless, these are some of the guidelines to follow and believed to arouse the interest of readers about mangroves.

3.7 SUCCESSION OF MANGROVES

The term 'succession' refers to a pool of individual plants growing as single stand or as communities forming complex structure on a land surface in a given time. Habitats of mangroves, which are considered as an ever-changing landscape at the interface of land and sea, are governed by some environmental factors such as tides, waves, river discharge, sea level movements and degree of salinity. Therefore, successional tendencies of mangroves are limited and restricted as well in particular locations. The action of environmental factors is the result of the occurrence of specific mangrove species on a particular land surface (Figure 3.4). In other way existence of specific mangroves species indicates what environmental/ecological factors govern a land surface they grow in. In mangroves habitats, different types of microalgae start assembling and initiate proliferation in newly silted and loose soil strata, followed by grass vegetation; both make the habitat rather consolidated mud. Usually, *Avicennia*, as a tree species, appears to be a pioneer in growing over mudflat surface through seeds germination. With times, mud flat zone becomes transformed into consolidated land, suitable for other mangroves to grow. Later consolidated zone becomes stable land and be ideal for major mangroves and mangroves associates to spread.

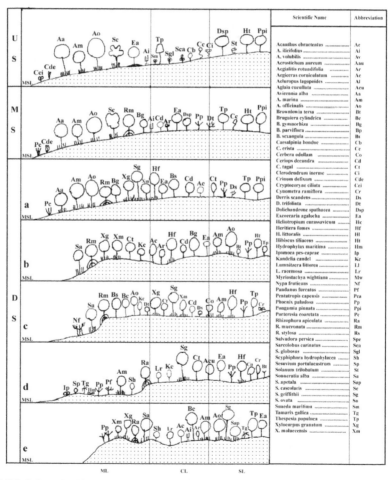

Scientific Name	Abbreviation
Acanthus ebracteatus	Ac
A. ilicifolius	Ai
A. volubilis	Av
Acrostichum aureum	Aau
Aegialitis rotundifolia	Ar
Aegiceras corniculatum	Ac
Aeluropus lagopoides	Al
Aglaia cucullata	Acu
Avicennia alba	Aa
A. marina	Am
A. officinalis	Ao
Brownlowia tersa	Bt
Bruguiera cylindrica	Bc
B. gymnorhiza	Bg
B. parviflora	Bp
B. sexangula	Bs
Caesalpinia bonduc	Cb
C. crista	Ce
Cerbera odollam	Co
Ceriops decandra	Cd
C. tagal	Ct
Clerodendrum inerme	Ci
Crinum defixum	Cde
Cryptocoryne ciliata	Cci
Cynometra ramiflora	Cr
Derris scandens	Ds
D. trifoliata	Dt
Dolichandrone spathacea	Dsp
Excoecaria agallocha	Ea
Heliotropium curassavicum	He
Heritiera fomes	Hf
H. littoralis	Hl
Hibiscus tiliaceus	Ht
Hydrophylax maritima	Hm
Ipomoea pes-caprae	Ip
Kandelia candel	Kc
Lumnitzera littorea	Ll
L. racemosa	Lr
Myriostachya wightiana	Mw
Nypa fruticans	Nf
Pandanus furcatus	Pf
Pentatropis capensis	Pca
Phoenix paludosa	Pp
Pongamia pinnata	Ppi
Porteresia coarctata	Pc
Rhizophora apiculata	Ra
R. mucronata	Rm
R. stylosa	Rs
Salvadora persica	Spe
S. globosus	Sgl
Scyphiphora hydrophylacea	Sh
Sesuvium portulacastrum	Sp
Solanum trilobatum	St
Sonneratia alba	Sa
S. apetala	Sap
S. caseolaris	Sc
S. griffithii	Sg
S. ovata	So
Suaeda maritima	Sm
Tamarix gallica	Tg
Thespesia populnea	Tp
Xylocarpus granatum	Xg
X. moluccensis	Xm

FIGURE 3.4 The diagram shows the distributional pattern of mangroves along intertidal zones in the estuaries of the Indian Sundarban: US, Upstream; MS, Midstream; DS, Downstream: a) mudflat, b) rivulet, c) small creek, d) sand dune and e) degraded land.

We mention here mangroves succession of the Indian Sundarban as the particular reference. The habitat falls under polyhaline zone with salinity varying between 18–30‰, particularly in the downstream. There may be a variety of succession patterns prevalent in a finer scale, based on the degree of salinity range with which tolerance of individual species varies. Even, there is different species combination with great variation within adjacent zone, which is poorly understood in ecological point of view. Nevertheless, a general succession pattern is visible with apparent vertical line from down to upstream along tidal ingression. On the other hand, in horizontal line mangroves communities succeed and exhibit great changes, based on land accretion and elevation characterized with maximum distance tidal range travels. *Porteresia coarctata,* as a grass species, first colonizes new silted up substrate stretching from US-MS (upper-mid stream) to DS (downstream) adjacent to estuary mouth. The appearance of *P. coarctata* indicates that this grass species can tolerate high salinity, even in the long duration of submergence during tidal ingression. *Nypa fruticans*, a mangrove palm may grow in such mudflat zone when it is fed with freshwater through its connection with the main flow of freshwater carrying river or channel of freshwater flow that reduces salinity. *N. fruticans* while occupying such substrate does not provide the space for *P. coarctata* to grow, a clear indication of species-specific succession on the basis of salinity. In silted up bed of mudflat zone, *Avicennia marina* and *A. alba* exhibit their distinct presence adjacent with *Porteresia coarctata* meadow and succeed until the land elevation occurs. In some areas with low salinity due to flushing of freshwater flow, a patch of *Sonneratia apetala* is visible in association with *Avicennia* spp. In such succession with slightly elevated land, *Heritiera fomes* may grow with scattered populations, although its presence is not so distinct as compared to other true mangroves. In consolidated bed with land elevation, the dominance of *Avicennia* spp. becomes weak and they are usually associated with other species, mainly *Rhizophora* spp., *Bruguiera* spp. and *Ceriops decandra,* members of family Rhizophoraceae. In some patches, *Kendelia candel* is dominant among the other members of Rhizophoraceae. However, in much-consolidated bed with slight elevated land, *Aegiceras corniculatum, Aegialitis rotundifolia, and Lumnitzera racemosa* are rather more visible plant communities in place of Rhizophoraceae members. When salinity becomes strong, *Scyphiphora hydrophylacea* occupies loosely consolidated bed, with its scattered populations. However, maximum number of species assemblage occurs in zones II, III and IV in the stretches from mudflat-consolidated land to consolidated-stable land surfaces (Table 3.2). *Xylocarpus* spp., *Avicennia officinalis*, *Ceriops decandra* and *Excoecaria agallocha* are prevalent among others in such zones. In succession of major mangroves follow a number of mangroves associates such as *Cynometra iripa, Aglaia cucullata,* and *Cerbera odollam,* the prominent species among others. In stable land with high elevation, *Excoecaria agallocha* and *Phoenix paludosa* are common among plant communities. *Ceriops decandra* forms scrubby thicket covering a large area, but in salt crusted area covered with *Suaeda maritima,* a high salt tolerating shrub as visible to form scattered scrub to spread over a substantial area. In the edge of stable land, dominance of mangroves associates such as *Thespesia populnea* and *Dolichandrone spathacea* are prevalent flora among others in association with back mangals.

On the other hand, a different scenario exists in upstream, with *Cryptocoryne ciliata* and *Crinum defixum,* herbaceous plants occupying rather a mature silted up substrate characterized with lower salinity as tidal ingression loses its strength to travel a farther distance. In consolidated substrate, *Acanthus ilicifolius* is very much distinct covering a large area with scattered patch, in association with *Clerodendrum inerme*, a bushy shrub.

3.8   MANGROVES ASSOCIATIONS

Aerial photographs or various other forms of remote sensing can interpret the species groupings, distribution and quantify their area coverage more accurately. Nevertheless, regular travels with scientific curiosity through mangroves forest enable workers to understand mangroves associations across intertidal zones, though delineating area coverage is impossible. Our study used to visit the Indian Sundarban and travel in and around forest area with the country boat in certain intervals. There may be a number of associations of mangroves vegetation, based on geomorphic, structural and functional features of mangrove environment on a micro scale. Most of the vegetation is with heterogeneity features. Species occur in various proportion and combination. All the mangroves associations are of some value in respect of particular zonation, but they usually fail to be considered as broad group of vegetation of the Sundarban in a landscape scenario.

TABLE 3.2 A General Trend on Succession of Major Mangroves, Mangroves Associates and Backs Mangals along Positions of Estuary as Well as the Intertidal Zones

Left margin vertical labels: **VERTICAL** and **RIVER**

Position of Estuary	Intertidal zonation				
	Zone -I (ML)	Zone -II (ML-CL)	Zone-III (CL)	Zone-IV (CL-SL)	Zone- V (SL)
US	•Aa •Am €Cci €Cde €Sb	•Aa ŧAl •Am •Ao •Sc •Ea	•Aa •Am •Ao •Sc €Hc •Ea	•Aa £Ai •Am •Ao •Ea •Sc €Sn •Dsp £Ci	§Cb §Cc •Di •Ht §Pca §St •Tp
US-MS	ŧPc •Aa •Am €Cci €Cde	•Aa •Am •Ao ŧMw •Ea •Sc	•Aa, €Ae •Am •Ao •Sc •Cd •Ea	•Aa £Ai •Am, •Ao •Cd •Ea €Sn €Sm ¥Pp §Sgl §Sca	§Ds §Dt •Tp •Ht §St
MS	ŧPc •Aa •Am •Ao	•Aa •Am •Ao ŧPc ŧMw €Sp €Sn •Rm •Bg •Ea	•Aa €Ae •Am •Ao •Rm •Cd £Ar £Ac •Sa •Bc •Bg •Ea	•Aa €Aau £Ai •Am •Ao •Cd •Ea •Acu ¥Pp •Ea	•Cr £Ppa
MS-DS	ŧ Pc ⌐Ip •Aa •Am	•Aa •Am •Ao ŧPc •Rm •Cd §Av •Kc •Ea	•Ra •Bg •Rm •Cd •Xg •Xm •Ct •Hf •Hl •Ll •Bc £Ar •Ea	£Ac •Bx •Cm •Ea •Ra •Rm ¥Pp €Spo £Bt	•Ea ¥Pp £Tg
DS	ŧPc ¥Np •Aa •Am •Sap ⌐Hm	•Aa •Am •Ao ¥Np •Sap £Sh •Lr •Ea	•Ra •Rm •Cd •Bc •Bg •Bp •Bx •Cd •Ct •Rs •Sg •So •Xg •Xm •Ea	£Ac £Ar •Aa •Am •Ao ¥Pp •Ea	•Ea ¥Pp £Tg
Estuary mouth	H O R I Z O N T A L				

◄———— Seaward   Landward ————►

Full form of species abbreviations are mentioned in Figure 3.4; • = Tree, £ = shrub, €= herb, ŧ = grass, ⌐ = creeper, § = twiner, ¥ = Palm; US = upstream, US-MS = upstream- midstream, MS = midstream, MS-DS = midstream-downstream, DS = downstream.

With experience through observation by travels, we present a structural classification of mangroves that relate to species association (Table 3.2) with five distinct groupings: (a) *Porteresia* association, (b) *Avicennia* association, (c) *Sonneratia* association, (d) Mixed species association, and (e) *Phoenix* association. Single species association usually means to have 80–100% density of population of a particular species in a given area, with which presence of other species, if any, seems to be irrelevant. In the Indian Sundarban, single species dominance is found in downstream estuary. On the other hand, no distinct mangroves associations are found in upstream (US), up-mid streams (US-MS), and midstream (MS) estuaries across the intertidal zones. Distinct mangroves associations exist from mid-down streams (MS-DS) and extend to downstream (DS) estuary. Evidently, *Sonneratia caseolaris* exhibits its dominant presence mostly in upstream estuary, with sparse vegetation; but it disappears from midstream estuary and no population recorded from downstream. On the contrary,

Excoecaria agallocha shows its diverse range of occupancy stretching from upstream to downstream estuary, with uniform tree height across the range of its distribution. Major mangroves occupy a wide range of intertidal areas covering zones I, II, III, IV, but disappear in zone V which is occupied with mangrove associates along with back mangals. However, both *P. paludosa* and *E. agallocha* exhibit their limited presence in zone V.

a. **Porteresia association:** A single stand vegetation is distinct at accretion land (mud flat) covering mostly zone I and extends to zone II. This vegetation occurs very densely and is inundated by all classes of tides. Ecological succession interprets that *Avicennia* seeds/seedlings are trapped at *Porteresia* coverage and initiates colonization of mangroves tree vegetation.

b. **Avicennia association:** *Avicennia alba* among three species, by and large, starts appearing at accretion land (mud flat) after *Porteresia* vegetation. Reproductive phenology interprets that since *Avicennia alba* among three species of *Avicennia* commences first flowering & fruiting, followed by *A. marina* and *A. officinalis* in order, so its seeds germination and seedlings development seems to be advance in succession compared to other two species. These three species mostly cover zone II, with dominant and dense vegetation coverage inundated by all classes of tides. However, the recent surveys observe that flowering period among three species, *A. alba, A. marina, A. officinalis* commences within the same period of the Indian Sundarban. In this context, the knowledge on phenological ecology of three species of *Avicennia* is little understood and needs to be explored further.

c. **Sonneratia association:** A single stand population of *Sonneratia apetala* is found dominating at the eastern part of the Indian Sundarban and initiates tree colonization at accretion land like *Avicennia* spp. It usually covers zones I and II; however, its population appears to be sparse.

d. **Mixed species association:** It comprises all the major mangroves, including shrubs and trees, covers a major area of the Indian Sundarban forest and seems dominating the main structure of mangroves forest, extending from zone II to zone IV. However, its presence is prevalent at zones III and IV. Aerial view of this structure exhibits wavy line stretching from middle of intertidal zone.

e. ***Phoenix* association:** *Phoenix paludosa* association constitutes
    one of the important and conspicuously dense vegetation of
    the Indian Sundarban. It is reported that the Royal Bengal tiger
    (*Panthera tigris tigris*) prefers hiding at *Phoenix* vegetation since
    its skin camouflages leaves of this palm vegetation. This single
    stand association spreads over a large-scale area, but usually domi-
    nates at zone IV.

Keeping aside the Indian Sundarban, Andaman & Nicobar Islands
exhibits luxuriant growth of mangroves tree, with dominant, dense and
distinct association of *Rhizophora* spp. A long stretch of *Rhizophora*
vegetation covers estuary shore, with a uniform height. Besides, *Nypa*
association is prevalent at small creeks of many Islets.

3.9   HABITATS AND DISTRIBUTION OF MANGROVES

With encounter of high tidal thrust in estuary, mangroves habitats exhibit
two conditions of landscape usually prevalent: (i) Accretion and (ii)
Degradation. Species distributions are distinct in accretion land and some
others in degradation zone. For instance, species of the genus *Avicennia*,
as pioneer tree species, are found growing mostly in accretion land; by
contrast, *Phoenix paludosa*, palm mangrove, mainly grows in a consolidated land characterized with degradation landscape. Both exhibit mono
specific stand in respective habitats. However, other mangroves exhibit
mixed vegetative forms towards consolidated zones, of which some
habitats are highly degraded. Since environmental factors are responsible
for the distribution of respective mangroves in the microhabitats, their
distributional range from seaward to landward vary greatly; some grow in
a wide range of distance, while others are restricted within limited range.
Mangroves growing in a greater distance are capable of tolerating wide
range of salinity and resistant to adversities of diverse habitats, compared
to those growing in limited range (Figure 3.5).

3.10   CLASSIFICATION OF MANGROVES VEGETATION FORMS

Mangroves distribution and distributional range (Figure 3.5) are related to
adaptive ability of respective species, governed by physical characteristics

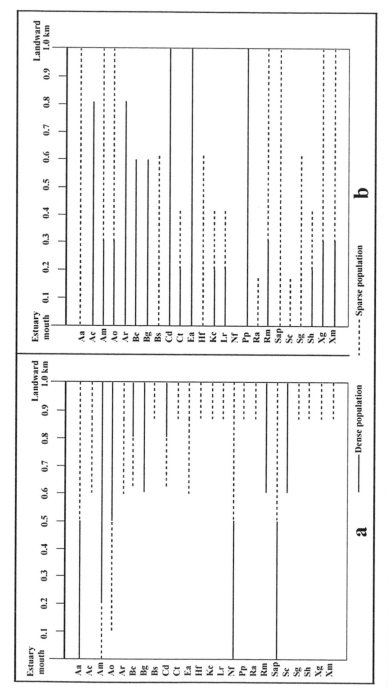

FIGURE 3.5 The diagram shows distributional range of mangroves within 1.0 km landward distance from estuary mouth in the Indian Sundarban, with conditions of microhabitats: (a) Accretion zone, and (b) Degradation zone.

of habitats, mostly hydrology, nutrient cycling and salinity. Therefore, diverse forms of mangroves vegetation exist worldwide, based on habitat conditions. Scientists identified these diverse forms and classified them with times as mentioned chronologically.

Lugo and Snedaker (1974): Six physiographic types of mangrove stands are considered, which are riverine mangroves, fringe mangroves, basin mangroves, overwash mangroves, scrub mangroves, and hammock mangroves.

I. Riverine mangroves exist along rivers and streams and are inundated twice a day. Gilmore and Snedaker (1993) considered those as riverine mangroves which occur on seasonal floodplains in areas in which natural patterns of freshwater discharge are consistent with an alternating cycle of high runoff/low salinity followed by low runoff/high salinity. Ewel et al. (1998) advocated that riverine mangroves were the most productive among mangrove communities due to its high nutrient concentrations associated with sediment trapping.

II. Fringe mangroves are those which exist along protected coastlines and islands and the exposed open waters of bays and lagoons. They are periodically inundated by tides and are vulnerable to erosion. They usually have well-developed root systems, which act as a barrier for protecting fragile coastlines.

III. Basin forests are those which are situated inland in depressions channelling terrestrial runoff towards the coast. They are irregularly flooded by tides and are vulnerable to flooding. They often serve as nutrient sinks for both natural and human-influenced ecosystem processes and have been as regular sources of wood products (Ewel et al. 1998).

IV. Overwash mangroves are those which are grown in subtidal to intertidal marine-dominated systems situated in isolated islands. They are flooded on each tidal cycle and their productivity is almost similar to fringe mangroves.

V. Scrub mangroves are those which are usually found in extreme environments. Their height is often limited caused by limiting nutrients and freshwater supply that may affect growth. Tidal inundation is infrequent.

VI. Hammock mangroves are those which occur on slightly elevated ground because of accumulation of organic peat over a depression. They experience irregular tidal flushing.

Cintron et al. (1985): All these mangroves, as identified by Lugo and Snedaker (1974), are grouped into three categories which are mainly found in the New World.

1. Rriverine mangroves;
2. Fringing (including overwash);
3. Basin (including dwarf and hammock).

Woodroffe (1990): There are mainly three types of mangroves based on dominant physical processes.

1. River dominated mangroves (sediment brought in by rivers);
2. Tide-dominated (sediment brought in by tides);
3. Carbonate settings (sediment is largely produced in situ, either as reef growth, calcareous sediment, or mangrove peat).

Ewel et al. (1998): This is a most recent classification which has simplified hybrid classification of habitats based on dominant physical processes: river dominated mangroves as riverine mangroves, tide-dominated mangroves as fringe mangroves, and interior mangroves as basin mangroves. Field workers may be able to identify roughly the specific category of habitats along with mangroves vegetation by dint of conditions of habitats and morphological appearance of mangroves. Below are mentioned features of different habitats;

I. Fringe mangroves:
   • Habitats are characterized with strong tidal thrust;
   • Stilt roots, buttress roots, knee roots, pneumatophores are common among trees;
   • Vegetation appears to be moderately dense;
   • Salinity is high.

II. Riverine mangroves:
   • Habitats are flooded by both river and tidal waters;
   • Vegetation appears to be dense;

- Trees appear to be luxuriant growth;
- Salinity is moderate.

III. Basin mangroves:
  - Habitats are located behind fringe and riverine vegetations, inundated occasionally;
  - Vegetation appears to be sparse;
  - Trees appear to be dwarf;
  - Salinity is very high due to evapotranspiration.

Field workers are unlikely to find all three kinds of mangroves vegetation in each Indian habitat. The Sundarban exhibits distinctly all three types of vegetation; almost similar pattern of three types of vegetation, but inconspicuous form, may be seen in Bhitarkanica. However, Godavari and Krishna deltas may exhibit both 'Fringe' and 'Basin' mangroves whereas 'Riverine' one is not distinct to see (Blasco and Aizpuru, 2002). Andaman & Nicobar Islands is of luxuriant mangroves vegetation, with dense coverage of canopy and distinct rooting pattern. On the contrary, 'Basin mangroves' are conspicuous in mangroves habitats of West Coast in India.

KEYWORDS

- **associations**
- **classification**
- **ecology**
- **position of estuary**
- **succession**
- **zonation of land surface**

CHAPTER 4

ADAPTATION AND PHENOLOGY OF MANGROVES

4.1 ADAPTATION

The term 'adaptation' defines the modification of organs or of whole organism to cope with the existing environmental conditions. Wide ranges of modifications are found in leaf, stem, root (vegetative organs) and reproductive organs of mangrove species. These morphological modifications of organs are distinctly visible in their habitats, even casual visitors can spot easily and differentiate them from other non-mangroves species growing adjacent to coastal regions. The modifications (Table 4.1) that mangroves adapt are essential to sustain in estuarine habitats characterized with a wide range of salinity and hence, the concentration of salt seems to be single determining factor of the mangroves environment. Mangroves are able to control the intake of salt and maintain water balance which is physiologically acceptable (Saenger, 2002).

4.2 SALT BALANCE AND MANGROVES

Excess salt that appears to be toxic in plant tissues needs to be removed prior to its intake in the metabolic process of mangroves; in the other words, mangroves require essentially to cope with high salinity through removal of such excess salt for survival (Scholander, 1968; Tomlinson, 1986). Popp (1995) has preferred using two terms 'tolerance' and 'resistance' in relation to salt and other stresses to mangroves instead only 'tolerance' used previously. In his argument, mangroves have developed resistance which indicates salt avoidance and regulation in one way, and salt tolerance or accommodation in other. All mangroves show some features of salt resistance that may broadly includes 'salt exclusion,' 'salt extrusion,' and 'succulence.'

4.2.1   SALT EXCLUSION

Roots act as a potential filter to exclude excess salt through effective filtration mechanism while taking up water entering into xylem sap (Popp et al., 1993). Genera which adapt such mechanism of salt exclusion include *Rhizophora, Ceriops, Bruguiera, Sonneratia, Excoecaria* and *Aegiceras*. Root systems are involved in exclusion of salt about 80–95% during water uptake, and the rest 5–20% salt entering into roots is considered essential for survival of mangroves (Saenger, 2002). No record of salt exclusion is, however, evident from leaves (Popp, 1995). Continuous salt exclusion by roots leads to localized build-up of high salinity around the absorption root zones of mangroves and in such case regular tidal inundation seems to wash out the build-up salt around the mangroves root zones (Saenger, 2002).

4.2.2   SALT EXTRUSION

Salt glands (Figure 4.1) in leaves are involved in salt extrusion. Genera which adapt such mechanism of salt extrusion include *Avicennia, Aegiceras, Aegialitis, Acanthus*, and *Rhizophora;* the last one extrudes salt through cork warts in leaves. The amount of salt extrusion varies among species in the range of 2.1–33%, with variable rates as recorded from 12–24 h (Popp, 1995; Weiper, 1995). Field et al. (1984) studied that a small slit-like opening develops between the cuticle of the leaf and that of the gland (Figure 4.1. iv–vi). Salt extrusion occurs through this silt.

4.2.3   SUCCULENCE

Leaves in mangroves that perform to storage salt are characterized with succulence. Saenger (2002) pointed out that succulence occurs to adjust accumulated salt through its dilution by increasing water content per leaf area volume, which is common strategy of halophytic adaptation. Succulence is a response primarily to the presence of Na^+ and Cl^-. Genera which adapt such mechanism of succulence include *Avicennia, Aegiceras, Acanthus, Sonneratia, Lumnitzera, Schyphiphora, Conocarpus and Rhizophora*. Succulence is attributed to increase of cell length in central mesophyll layers as reflected in anatomical structure, with moisture content in the range of 65–70% (Saenger, 2002).

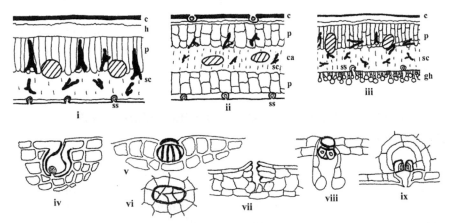

FIGURE 4.1 Leaf anatomy, diagrammatic sectional view: (i) *Rhizophora mucronata*: dorsiventral leaf with (*c*) thick cuticle, (h) hypodermis, (p) palisade tissue, (sc) sclerides and (ss) sunken stomata; (ii) *Sonneratia caseolaris*: Isobilateral leaf with (c) thick cuticle, (p) palisade tissue, (ca) central aqueous tissue, (sc) sclerides and (ss) sunken stomata; (iii) *Heritiera fomes*: dorsiventral leaf with (c) thin cuticle, (p) palisade tissue, (sc) sclerides, (ss) sunken stomata and (gh) glandular hairs, (iv) *Aegiceras corniculatum*: T. S. of salt gland in adaxial surface, (v–vi) *Aegialitis rotundifolia*: T. S. of salt gland with side view and top view in adaxial surface, (vii) *Excoecaria agallocha*: T. S. of lenticels; (viii) *Avicennia marina*: T. S. of stomata with glandular hair in abaxial surface, (ix) *Phoenix paludosa*: T. S. of sunken stomata in abaxial surface.

4.3   LEAF AND STEM SYSTEMS AND ADAPTATIONS

Mangroves leaves exhibit some modifications to cope with high salinity in habitats, which are distinctly visible in field (Figures 4.1 and 4.2): thick cuticle with waxy coating on upper surface in fleshy or succulent or thick leaves, and dense hairs with velvet like appearance in lower surface in coriaceous leaves. These modifications help respective species conserve water which otherwise get lost due to evaporation or transpiration. In saline environment water availability in form of cell sap needs to be stored and protected, because quantity of salt in form of solute remains as much higher amount as compared to the available water as solvent. So, all these modifications are related to conservation as well as protection of water essential for physiological function.

In stem, lenticels (Figure 4.1. vii) and galls are common features as seen on stem bark as part of a strategic adaptation. These modifications facilitate the breathing of cells through inhaling oxygen from air in one way and take part in excretion in other. Usually, the amount of oxygen in cell sap remains insufficient quantity due to high salinity in habitats, which is substantiated by the process of inhaling from air through lenticels and galls.

4.4 WATER CONSERVATION AND XEROMORPHY OF MANGROVES

Water conservation is an adaptive response of mangroves as an essential strategy to cope with saline environment and for which leaves of the most mangroves exhibit a range of xeromorphic features. Below are mentioned some features related to xeromorphy of mangroves (Saenger, 1982; Tomlinson, 1986; Naskar and Mandal, 1999; Saenger, 2002; Mandal and Naskar, 2008).

a. Thick walled and often multi-layered epidermis present on the upper leaf surface, covered with a thick, waxy, lamellar cuticle (Figure 4.1. i & ii) that seem to reduce evaporative loss.
b. A well-developed hypodermis seems to be conspicuous as xeromorphic feature in relation to various functions such as water storage, salt accumulation or osmoregulation, mesophyll protection through heat dissipation, and nutrients conservation (Figure 4.1. i).
c. Dense layer of various shaped hairs present at lower surface of leaves (Figure 4.1. iii) seems to cover salt glands and stomata to check the loss of water through these apertures.
d. Distribution of cutinized and sclerenchymatous cells (Figure 4.1. i, ii & iii) found throughout the leaf is a part of xeromorphic features that respond to physiological dryness of saline environment, with provision of mechanical support against frequently occurrence of natural calamities including strong wind and cyclone.
e. Dominance of palisade mesophyll (Figure 4.1. i, ii & iii), at the expanse of spongy mesophyll, is a characteristic feature of xeromorphy, which provides toughness and rigidity to the leaves and reduces damage from wilting, with provision of water conservation.

f.   Stomata sunk beneath the level of epidermis (Figure 4.1. viii & ix) in several mangroves seem to facilitate water storage in one hand, and to check the evaporative loss of water in other.

g.   Have more vessels per unit area with distinctly smaller pores observed in wood, with a larger cross-sectional area, than do their nearest inland relatives or even the same mangroves species while growing under less saline conditions (Panshin, 1932; Janssonius, 1950; Vliet 1979; Naskar and Mandal, 1999).

The above features are difficult to study in field without anatomical dissection of leaves or stem of respective species. However, mangroves with such xerophytic features are distinctly visible in their morphological expressions, particularly in leaves (Figure 4.2). Field workers may be able to find all these features through leaf morphology with their keen observation and curious interest. For instance, mangroves exposed continuously to high salinity exhibit higher thickness of leaf than the same species has lower leaf thickness while growing in low salinity (Saenger, 2002).

4.5   ROOT SYSTEM AND ADAPTATIONS

Muddy substrate of estuarine areas are characterized, by and large, with anaerobic condition due to scanty of soil porosity. The anaerobic condition seems to be extreme when water logging persists. Estuarine soil is, thus, devoid of sufficient amount of oxygen essential for physiological function of under-ground and related organs. Reasonably, soil of mangrove habitats is known as 'Physiologically dry.' Besides, estuarine ecosystems encounter natural phenomena like sea surges, high tidal flow, strong wind flow, soil erosion, coastal storms including cyclones, hurricanes, etc. Mangroves develop a survival strategy in root system to withstand against such adversities. This survival strategy is referred to 'halophytic adaptation,' responsible for a variety of modification in roots across mangroves species, with provision of both mechanical support and physiological function. In field, there are two kinds of rooting pattern distinct in mangroves: (i) underground root system and (ii) aboveground root system. Concisely the modification of roots essential to survival of mangroves is discussed.

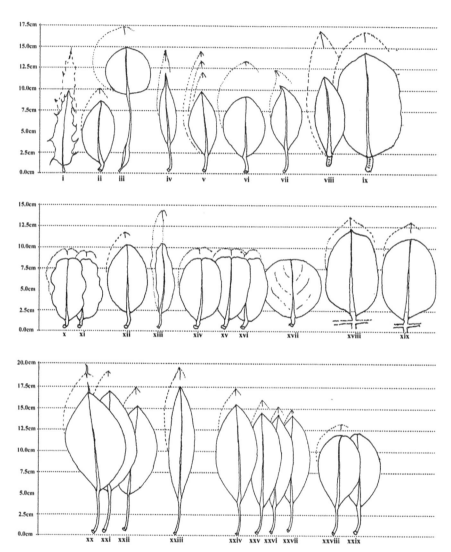

FIGURE 4.2   The diagram shows shape and range of leaf size of mangroves: i. *Acanthus ilicifolius*, ii. *Aegiceras corniculatum*, iii. *Aegialitis rotundifolia*, iv. *Avicennia alba*, v. *A. marina*, vi. *A. officinalis*, vii. *Excoecaria agallocha*, viii. *Heritiera fomes*, ix. *H. littoralis*, x. *Lumnitzera racemosa*, xi. *L. littorea*, xii. *Scyphiphora hydrophylacea*, xiii. *Sonneratia apetala*, xiv. *S. caseolaris*, xv. *S. griffithii*, xvi. *S. alba*, xvii. *S. ovata*, xviii. *Xylocarpus granatum*, xix. *X. moluccensis*, xx. *Rhizophora mucronata*, xxi. *R. apiculata*, xxii. *R. stylosa*, xxiii. *Kandelia candel*, xxiv. *Bruguiera gymnorhiza*, xxv. *B. sexangula*, xxvi. *B. cylindrica*, xxvii. *B. parviflora*, xxviii. *Ceriops decandra*, xxix. *C. tagal*.

4.5.1 UNDERGROUND ROOT SYSTEM

Major mangroves exhibit a horizontally spreading **cable root** system just beneath the soil surface (sometimes exposed above the ground), with descending roots known as **anchor roots** at certain intervals. In addition of **tap roots**, comparatively smaller fibrous like roots known as **nutritive roots** develop from anchor roots. All these roots may be considered to constitute underground root system (Figure 4.3i).

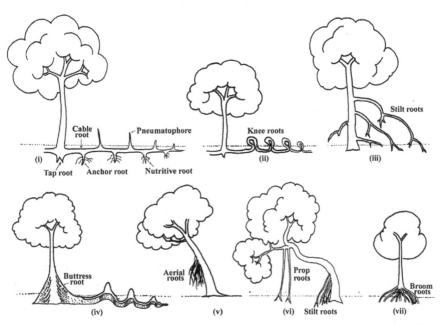

FIGURE 4.3   The diagram shows a wide range of root system in mangroves: (i) variety of underground and aboveground root systems; (ii) knee roots; (iii) stilt roots; (iv) buttress root with extended plank; (v) aerial roots; (vi) prop roots and stilt roots; (vii) broom roots.

4.5.2 ABOVEGROUND ROOT SYSTEM

A wide range of root diversity is prevalent in most mangroves species. **Pneumatophores** (Figure 4.3i): Roots arise, at certain intervals, from cable roots and ascend upward into the air against gravitation as in *Avicennia alba, A. marina, A. officinalis, Sonneratia alba, S. apetala, S. caseolaris, S. griffithii, S. ovata, Xylocarpus granatum, X. moluccensis,*

Heritiera fomes. However, upward roots arising against gravitation are known as **Pneumatothods** in *Phoenix paludosa,* though such roots are found rarely and obviously in nonporous consolidated substrate. Pneumatophores of some mangroves exhibit lenticels (*Avicennia*) and bark removal (*Sonneratia*) with maturity, which is related to certain physiological functions.

Knee roots (Figure 4.3ii):Some roots develop from cable roots, rising upward above the surface and again bending downward as seen in *Bruguiera* spp. **Stilt roots** (Figure 4.3iii, vi): A number of roots develop around from trunk base at certain height above the ground, descending with numerous branches and penetrating into ground as in species of *Rhizophora* and *Acanthus ilicifolius.* Stilt roots, sometimes, appear to be so dense that stem base becomes inconspicuous to be visible and root surface characterized with enormous distinct lenticels. **Buttress roots** (Figure 4.3iv): Roots develop from trunk base like stilt roots, but found to be flattened, blade like structure as in species of *Xylocarpus* and *Heritiera.* Buttresses, sometimes in species of both the genera, extend away from trunk base like ribbon structure known as **plank buttress** (Figure 4.3iv). **Aerial roots** (Figure 4.3v): Roots develop from branches or trunk, descending towards ground, but not penetrating into substrate as in species of *Avicennia, Sonneratia* and *Rhizophora.* Besides, the present authors observed other two types of above ground roots which are similar to the origin of stilt roots, but appear differently. **Prop roots** (Figure 4.3vi): Roots develop from bending branches in rare occasion and penetrating into ground as seen in *Rhizophora mucronata.* **Broom roots** (Figure 4.3vii):Roots develop from trunk base immediately above the ground and forming broom like structure like fibrous appearance as seen in *Bruguiera gymnorhiza, Kandelia candel, Ceriops decandra and C. tagal.*

All these modified roots have two different functions like mechanical support and physiological activity in respect of saline environment, but are mangrove species-specific (Table 4.1). Mechanical support is essentially required to the entire plant or part of plant or trunk base against natural adversities such as soil erosion, sea surges, high tidal flow, strong wind, and sea waves. The degree of mechanical support varies depending on the environmental condition of respective habitats. Physiological activity mainly includes breathing through inhaling oxygen from air, apart from additional functions as photosynthesis and excretion in few species.

TABLE 4.1 Modifications of Different Organs in Mangrove Species and Their Respective Functions

Organs	Attribute	Appearance	Function		Species
			Mechanical	Physiological	
Leaves	Thick cuticle	Glossy	—	To reduce water loss through transpiration by reflecting sunlight	Aegialitis rotundifolia, Bruguiera cylindrica, B. gymnorhiza, B. parviflora, B. sexangula, Ceriops decandra, C. tagal, Excoecaria agallocha, Kandelia candel, Rhizophora apiculata, R. mucronata, R. stylosa
	Gregarious hair (Capitate hairs on leaf ventral surface of Avicennia. Lepidote in case of H. fomes)	Velvet	—	To reduce water loss through transpiration by covering stomata	Avicennia alba, A. marina, A. officinalis, Heritiera fomes, H. littoralis
	Succulence	Fleshy	—	To accumulate and store excess salt	Sonneratia alba, S. apetala, S. caseolaris, S. griffithii, S. ovata, Aegiceras corniculatum, Lumnitzera littorea, L. racemosa, Scyphiphora hydrophylacea
Stems	Galls	Out growth	—	To breathe and excrete excess salt	Excoecaria agallocha
	Lenticels	Enormous pores		To breathe and excrete excess salt	Avicennia alba, A. marina, A. officinalis, Excoecaria agallocha, Sonneratia alba, S. apetala, S. caseolaris, S. griffithii, S. ovata
Root (under ground)	Cable roots	Long horizontal	To support trunk		Common features of the mangroves which have pneumatophores and aerial roots.

TABLE 4.1 *(Continued)*

Organs	Attribute	Appearance	Function — Mechanical	Function — Physiological	Species
	Anchor roots	Descending from cable root	To support cable root		
	Nutritive roots	Fibrous		To accumulate nutrients	
	Tap root	Main anchoring root	To support entire canopy		
Roots (above ground)	Pneumatophores (anti gravitational)	i. Pencil ii. Rod iii. Buffalo horn iv. Flat conical		a. To breathe and excrete excess salt, along with photosynthesis b. To excrete through removal of peeling	*a.* (i) *Avicennia alba, A. marina, A. officinalis* ii. *Sonneratia alba, S. apetala, S. caseolaris, S. griffithii, S. ovata* iii. *Xylocarpus granatum, X. moluccensis* iv. *Heritiera fomes, H. littoralis* *b. Sonneratia alba, S. apetala, S. caseolaris, S. griffithii, S. ovata*
	Pneumatothods (anti gravitational)	Pencil		To breathe	*Phoenix paludosa*
	Knee roots	Bending Knee	To support trunk base	To breathe	*Bruguiera gymnorhiza*
	Stilt roots	Bow	To support trunk base	To breathe	*Rhizophora apiculata, R. mucronata, R. stylosa*

TABLE 4.1 (Continued)

Organs	Attribute	Appearance	Function Mechanical	Function Physiological	Species
	Buttress roots (Plank roots)	Flat extended from trunk base	To support trunk base		*Xylocarpus granatum, X. moluccensis, Heritiera fomes, H. littoralis*
	Aerial roots	Hanging downward		To breathe	*Avicennia marina, A. alba, A. officinalis*
	Prop roots	Straight downward	To support bending branch	To breathe	*Rhizophora apiculata*
	Broom roots	Broom	To support trunk base	To breathe	*Bruguiera gymnorhiza, C. tagal, Kandelia candel*
Reproductive organs	Vivipary	Hanging rod		To have secure germination	*Bruguiera cylindrica, B. gymnorhiza, B. parviflora, B. sexangula, Ceriops decandra, C. tagal, Kandelia candel, Rhizophora apiculata, R. mucronata, R. stylosa*
	Crypto vivipary	Hidden hypocotyl		To have secure germination	*Avicennia marina, A. alba, A. officinalis; Aegiceras corniculatum, Aegialitis rotundifolia, Nypa fruticans*

4.6   REPRODUCTIVE ORGANS AND ADAPTATION

Mangroves increase progenies by sexual reproduction, except species of *Acrosticum*. Fruit acts as an exclusive reproductive organ, which bears seeds to germinate new propagules. Initially, development of seeds germination in mangroves is very much sensitive to high salinity and that may become withered, until suitable substrate is present. To avoid such high salinity seeds germinate without any dormant period and hypocotyl comes out while fruits remain attached with branches. This phenomenon is referred to viviparous germination as seen in species of *Rhizophora, Bruguiera, Ceriops* and *Kandelia* – all belong to the family Rhizophoraceae. The fruit exhibits hypocotyl distinctly visible while being attached with mother plant and until it drops down substrate. Another form of germination is referred to cryptovivipary as seen in the species of *Avicennia, Aegialitis, Aegiceras* and *Nypa*. In cryptovivipary, seed germinates to develop radicle which has restricted elongation and remains hidden within fruit coat.

4.7   DISPERSAL OF PROPAGULES

Mangroves require establishing their progenies through dispersal of reproductive units, referred to propagules which vary among respective mangroves: spores in species of *Acrostichum*, a single seed in species of *Excoecaria* and *Phoenix*, a one-seeded fruit in *Cynometra*, several seeded fruit in species of *Sonneratia* and *Xylocarpus*, a multiple fruit in *Heritiera*, an aggregated fruit in *Nypa*, and fruit carrying hidden propagule in species of *Avicennia, Aegiceras* and *Aegialitis* and conspicuous propagules (Figure 4.4) in species of *Bruguiera, Ceriops, Kandelia* and *Rhizophora*. A common character all the mangroves possess is buoyancy prevalent in all types of propagules. Buoyancy occurs in a various parts of fruit that may vary in respective species. For example, hypocotyl performs buoyancy in species of *Rhizophora*, pericarp in species of *Avicennia*, endosperm in species of *Xylocarpus*, the seed testa in species of *Nypa* and of *Sonneratia*, and seed in species of *Acanthus*. Saenger (2002) reported that distance of dispersal ability of mangroves varied greatly among species: propagules may travel in the range of 250–350 km under different conditions, with expanses of 300 days as instance of maximum time recorded in *Rhizophora* (Rabinowitz, 1978).

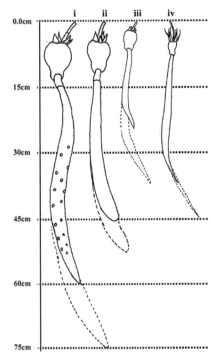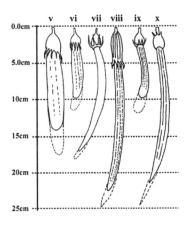

FIGURE 4.4   The diagram shows shape and range of hypocotyl lengths in mangroves: (i) *Rhizophora mucronata*, (ii) *R. apiculata*, (iii) *R. stylosa*, (iv) *Kandelia candel*, (v) *Bruguiera gymnorhiza*, (vi) *B. sexangula*, (vii) *B. cylindrica*, (viii) *B. parviflora*, (ix) *Ceriops decandra* and (x) *C. tagal*.

4.8   REPRODUCTIVE PHENOLOGY

In most of the mangroves, flowering commences at pre-summer months, followed by fruiting that occurs during peak monsoon months (Table 4.2). The period of flowering and fruiting in mangroves is mentioned here, along with respective months that relate to the climatic conditions of the Indian subcontinent. Reproductive phenology is considered an important criterion for authenticity of proper identification. And identification of species through flowering morphology is most reliable method that benefits all in the field study, particularly those intended to undertake restoration activities of mangroves. Shifting of phenological time of specific mangrove or of all, if happens due to climatic changes, may be recorded as an important evidence for further study.

TABLE 4.2 Monthly Occurrence of Flowering and Fruiting of Mangroves

Scientific Name	Months (January–December)											
	J	F	M	A	M	J	J	A	S	O	N	D
Acanthus ebracteatus	⋮	⋮	⋮	⋮	*Ô	*Ô	*Ô	*Ô	⋮	⋮	⋮	⋮
Acanthus ilicifolius	·	⋮	⋮	⋮	*Ô	*Ô	*Ô	*Ô	⋮	⋮	⋮	⋮
Acanthus volubilis	⋮	⋮	⋮	⋮	⋮	*Ô	*Ô	*Ô	*Ô	⋮	⋮	⋮
Acrostichum aureum	ö	ö	⋮	⋮	⋮	⋮	⋮	⋮	⋮	⋮	ö	ö
Aegialitis rotundifolia	⋮	⋮	⋮	*	*	*Ô	*Ô	*Ô	O	O	⋮	⋮
Aegiceras corniculatum	*	*	*	*Ô	*Ô	O	O	O	O	⋮	⋮	⋮
Aglaia cucullata	*Ô	⋮	⋮	⋮	*Ô	*Ô	⋮	⋮	⋮	*	*Ô	*Ô
Avicennia alba	⋮	⋮	*	*	*Ô	*Ô	O	O	O	O	⋮	⋮
Avicennia marina	⋮	⋮	⋮	*	*	*Ô	*Ô	O	O	O	O	⋮
Avicennia officinalis	⋮	⋮	⋮	*	*	*	*	*Ô	O	O	⋮	⋮
Brownlowia tersa	⋮	⋮	⋮	⋮	⋮	*	*Ô	*Ô	*Ô	⋮	⋮	⋮
Bruguiera cylindrica	O	O	*Ô	*Ô	*Ô	*Ô	*Ô	*Ô	*Ô	*Ô	*Ô	O
Bruguiera gymnorhiza	O	O	*Ô	*Ô	*Ô	*Ô	*Ô	*Ô	*Ô	*Ô	O	O
Bruguiera parviflora	O	O	*Ô	*Ô	*Ô	*Ô	*Ô	*Ô	*Ô	*Ô	O	O
Bruguiera sexangula	O	O	*Ô	*Ô	*Ô	*Ô	*Ô	*Ô	*Ô	*Ô	*Ô	O
Caesalpinia bonduc	O	O	⋮	⋮	⋮	⋮	*	*	*	*Ô	⋮	O
Caesalpinia crista	*	*Ô	*Ô	O	O	O	O	⋮	⋮	⋮	*	*
Cerbera odollam	⋮	⋮	*	*	*	O	⋮	⋮	O	⋮	⋮	⋮
Ceriops decandra	O	*Ô	*Ô	*Ô	*Ô	*Ô	*Ô	*Ô	*Ô	O	O	O
Ceriops tagal	O	O	O	*Ô	*Ô	*Ô	*Ô	*Ô	*Ô	O	O	O

TABLE 4.2 (Continued)

Scientific Name	Months (January–December)											
	J	F	M	A	M	J	J	A	S	O	N	D
Clerodendrum inerme	⋯	⋯	⋯	*	*	*	*○	*○	○	○	○	⋯
Crinum defixum	⋯	⋯	⋯	⋯	⋯	*○	*○	*○	*○	⋯	⋯	⋯
Cynometra ramiflora	*○	*○	*○	⋯	⋯	⋯	⋯	⋯	*○	*○	*○	*○
Derris scandens	⋯	⋯	⋯	⋯	⋯	⋯	○	○	*	○	○	○
Derris trifoliata	⋯	⋯	⋯	*	*	*	*	*	⋯	⋯	*	*
Dolichandrone spathacea	*○	○	○	⋯	⋯	⋯	⋯	⋯	⋯	⋯	⋯	⋯
Excoecaria agallocha	⋯	*	*	*	*○	*○	○	○	*	*	○	○
Heliotropium curassavicum	⋯	⋯	⋯	⋯	⋯	⋯	⋯	⋯	*	*	*	⋯
Heritiera fomes	⋯	⋯	⋯	*	*	*	○	○	○	○	○	⋯
Heritiera littoralis	⋯	⋯	⋯	⋯	*	*	*	○	○	○	○	⋯
Hibiscus tiliaceus	*○	*○	*	*	*	*	*	*	*	*	*	*
Hydrophylax maritima	⋯	⋯	⋯	⋯	⋯	⋯	*	*	○	⋯	⋯	⋯
Ipomoea pes-caprae	*○	*○	○	*○	*○	*○	*○	*○	*○	*	*○	*○
Kandelia candel	⋯	⋯	⋯	*	○	○	○	○	○	○	○	○
Lumnitzera littorea	⋯	⋯	⋯	*	⋯	*	*	*	*	○	○	⋯
Lumnitzera racemosa	⋯	⋯	⋯	*	*	*	○	○	○	○	○	○
Nypa fruticans	○	○	○	*○	*○	*○	*○	*○	*○	*○	*○	*○
Pongamia pinnata	⋯	⋯	⋯	*	*	○	○	○	⋯	⋯	⋯	⋯
Phoenix paludosa	⋯	⋯	*	*	*	*	⋯	⋯	⋯	⋯	⋯	⋯
Porteresia coarctata	⋯	⋯	⋯	⋯	⋯	*	*	*	*○	○	○	○

TABLE 4.2 *(Continued)*

Scientific Name	Months (January–December)											
	J	F	M	A	M	J	J	A	S	O	N	D
Rhizophora apiculata	Ô	Ô	*Ô	*Ô	*Ô	*Ô	*Ô	*Ô	*Ô	*Ô	Ô	Ô
Rhizophora mucronata	Ô	*Ô	*Ô	*Ô	*Ô	*Ô	*Ô	*Ô	*Ô	*Ô	Ô	Ô
Rhizophora stylosa	Ô	*Ô	*Ô	*Ô	*Ô	*Ô	*Ô	*Ô	Ô	*Ô	Ô	Ô
Sarcolobus carinatus	…	…	…	…	…	…	*	*	Ô	Ô	…	…
Sarcolobus globosus	Ô	…	…	…	…	…	…	…	*	*	*	Ô
Scyphiphora hydrophylacea	…	…	…	*	*Ô	*Ô	*Ô	Ô	Ô	…	…	Ô
Sesuvium portulacastrum	Ô	…	…	…	*Ô	*Ô	*Ô	…	…	*	*	*Ô
Sonneratia alba	…	…	…	*Ô	*Ô	*Ô	*Ô	…	Ô	…	…	…
Sonneratia apetala	…	…	…	…	*	*	*	…	Ô	…	…	…
Sonneratia caseolaris	…	…	*	*Ô	*Ô	*Ô	*Ô	*Ô	*Ô	*Ô	…	…
Sonneratia griffithii	…	…	…	*Ô	*Ô	*Ô	*Ô	*Ô	*Ô	*Ô	…	…
Sonneratia ovata	…	…	…	…	*Ô	*Ô	*Ô	*Ô	*Ô	*Ô	…	…
Suaeda maritima	…	…	…	…	…	…	…	…	*	*	*Ô	Ô
Tamarix gallica	*Ô	*Ô	*Ô	…	…	…	…	…	…	…	*	*Ô
Thespesia populnea	*Ô	*Ô	*Ô	*Ô	*Ô	*Ô	*Ô	*Ô	*Ô	*Ô	*Ô	*Ô
Xylocarpus granatum	…	…	…	*	*Ô	*Ô	Ô	Ô	…	…	…	…
Xylocarpus moluccensis	…	…	*	*	*Ô	Ô	Ô	Ô	Ô	Ô	Ô	Ô

*Denotes flowering time, Ô fruiting time, *Ô flowering and fruiting times together, and ô spore formation.

4.8.1 STEPS TO STUDY FLOWER

- Have information about the flowering period of respective mangroves.
- Choose the middle time of flowering period to get a maximum number of full-grown flowers.
- Go through the details or salient features of flower mentioned in Chapter 5.
- Note all the attributes and find whether they are matched with the flowers found in the field.

4.8.2 STEPS TO STUDY FRUIT

- Have information about the fruiting period of respective mangroves.
- Have fruit shape and size of respective mangroves mentioned in Chapter 5.
- Have information about the time gap between flowering and fruiting because flowering and fruiting occurs at the same time in many species.

A great variation of morphological features, as well as adaptive behaviors, occurs across the mangroves species. Every species has its own way of adaptation, even individuals of same species may exhibit variable features adapted to cope with the condition of particular habitat. For instance, individuals of *Rhizophora mucronata* may produce enormous number of stilt roots in the habitat affected with the harsh environmental condition due to strong tidal current, compared to very less number of stilt roots developed in the individuals of the same species while growing in the habitat inundated with low tidal fluctuations. A wide range of modified organs are prevalent in mangroves communities that signify variation of adaptation as per the need of individuals and modification of organs occurs accordingly – which are distinct as well as visible. Field workers are suggested to observe carefully different attributes of species to study, not to stick one feature while identifying the targeted species.

KEYWORDS

- adaptation
- propagules
- root systems
- salt balance
- water conservation
- xeromorphy

CHAPTER 5

CLASSIFICATION AND IDENTIFICATION OF MANGROVES

5.1 CLASSIFICATION

Classification simply means to group a number of things (living or non-living) based on similarities or dissimilarities, for easy identification, explanation, and documentation. Different species of mangroves are classified here to make them as broad groups and segregate from one group to another, mainly between the communities, for identification in the landscape vegetation. Few attributes, which characterize major mangroves, are visible in the spot, distinctive among respective communities and are considered as the gateway of readers for easy identification of mangroves communities in the field. This classification is only for major mangroves, the main component of mangroves vegetation and for which the mangrove ecosystems recognized worldwide.

5.1.1 BROAD GROUPS

Group I. Viviparous hypocotyl (seed germinates while fruit remains attached with mother plant) *Bruguiera* (*B. cylindrica, B. gymnorhiza, B. parviflora, B. sexangula*)
 Ceriops (*C. decandra, C. tagal*)
 Kandelia candel
 Rhizophora (*R. apiculata, R. mucronata, R. stylosa*)

Group II. Crypto-vivipary ('Crypto' means hidden. The attribute is not visible in the field; interested ones, however, need little effort to look it within fruit coat of respective species. This germination is also known as 'incipient' vivipary)
 Acanthus (*A. ebracteatus, A. ilicifolius, A. volubilis*)

Aegialitis rotundifolia
Aegiceras corniculatum
Nypa fruticans

Group III. Pneumatophore (Root arising above ground against gravitation)
Avicennia (*A. alba, A. marina, A. officinalis*)
Heritiera fomes
Phoenix paludosa (occasional; here this roots known as pneumatothod)
Sonneratia (*S. alba, S. apetala, S. caseolaris, S. griffithii, S. ovata*)
Xylocarpus (*X. granatum, X. moluccensis)*

Group IV. Aerial roots (Roots developing from branches or trunk or trunk base, moving down the ground)
Avicennia (*A. alba, A. marina, A. officinalis*) – occasional
B. gymnorhiza
Rhizophora (*R. apiculata, R. mucronata, R. stylosa*)

Group V. Above ground roots (Roots developing from trunk base)
Aegialitis rotundifolia
Excoecaria agallocha
Heritiera (*H. fomes, H. littoralis*)
Lumnitzera (*L. racemosa, L. littorea*)
Scyphiphora hydrophylacea
Xylocarpus (*X. granatum, X. moluccensis*)

5.1.2 DISTINCT GROUPS

Having acquainted with the attributes of broad groups of mangroves, the readers need to proceed further to have as much as knowledge about them at the generic level.

Several steps of the distinct groups are mentioned below. Not that every step is required to identify all the species, but that each step representing the attributes of the respective genus needs to be matched while identifying species in the field. However, very few species belonging to mangroves have their morphological attributes not so distinctly visible for their identification in the field.

Step 1: View the canopy, if there are viviparous hypocotyls hanging from branches; consider all belonging to Group I and measure the length of mature hypocotyls, by assumption or using measuring tools.

Step 1a. Hypocotyl in the range of 8–25 cm, hanging from medium to large tree: *Bruguiera*

Step 1b. Hypocotyl in the range of 8–25 cm, hanging from shrub/small tree: *Ceriops*

Step 1c. Hypocotyl in the range of 40–50 cm with smooth surface and pointed end: *Kandelia*

Step 1d. Hypocotyl in the range of 20–75 cm, with or without warty surface: *Rizophora*

Step 2: Rupture fruit coat, if there is germinated seed lying within it, consider all belonging to Group II.

Step 2a. Germinated seed lying into globose-eliptic fruit: *Acanthus*

Step 2b. Germinated seed lying into narrow-cylindrical fruit: *Aegialitis*

Step 2c. Germinated seed lying into fruit with slightly bending end: *Aegiceras*

Step 2d. Germinated seed arising upright, looks a small coconut fruit: *Nypa*

Step 3: Look at ground or base of the vegetative stand, if there are pnematophores arising above the ground against gravitation; consider all belonging to the group III.

Step 3a. Pneumatophores look wood pencil structure, rising upward with numerous numbers on the ground: *Avicennia*

Step 3b. Pneumatophores look flatten with blunt end: *Heritiera*

Step 3c. Pneumatothods look wood pencil structure, with regular rings encircling the periphery, (pnematothods found occasionally): *Phoenix*

Step 3d. Pneumatophores look cylindrical rod structure, with peeling layers on bark: *Sonneratia*

Step 3e. Pneumatophore look buffalo horn/peg-like structure, stout with almost pointed end: *Xylocarpus*

Step 4: View the trunk or branches of mangroves, if there are roots developing and moving down or hanging; consider them belonging to Group IV.

Step 4a: Roots appear hanging from trunk/branches occasionally: *Avicennia*

Step 4b: Roots arise above, look knee shape and spread: *Bruguiera*

Step 4c: Roots look bow/ slanting architecture arising densely from trunk: *Rhizophora*

Step 4d: Roots appear hanging from branches occasionally: *Rhizophora*

Step 5: View the trunk base, if there are roots spreading horizontally on ground; consider them belonging to Group V.

Step 5a: Roots, fused with trunk base, look swollen like conical shape: *Aegialitis*

Step 5b. Roots look serpentine spreading horizontally: *Excoecaria*

Step 5c. Roots look flat slanting as a large buttress and extend away as plank buttress from the trunk base: *Heritiera*

Step 5d: Roots look wavy spreading horizontally: *Lumnitzera*

Step 5e: Roots look slightly wavy and exposed: *Schyphiphora*

Step 5f. Roots look flat slanting as small buttress and extend away as plank buttress from the trunk base: *Xylocarpus*

5.2 IDENTIFICATION OF SPECIES

Identification is best known a process of showing, providing or recognizing what the thing is. There may be several techniques by which a thing may be known to somebody, with clear understanding. Here, the attributes of mangroves species are mentioned, with five ways such as (i) diagnostic description, (ii) sketches, (iii) images, (iv) salient morphometry, and (iv) key characters. The attributes of a species, presented in such a way, are believed to ease those who need identification of respective species and also distinguish one to another.

Features of mangroves, thus, include five ways to follow:

a. Go through diagnostic description, a glimpse of characters to provide information about the species.

b. See through sketches and images of the species to understand and match whether the species really is.

c. Verify the species with its salient morphometry, if someone gets its accuracy.

d. Be familiar with its characters of spot identifications to distinguish it from another one.

e. Have a chance to easily distinguish closely related species belonging to the same genus; for there is a comparative assessment of attributes among related species.

The method of identification may be useful in both field and laboratory, even in museum for herbarium specimen.

5.3 MAJOR MANGROVES

Major mangroves include 32 species arranged alphabetically, with each species described concisely.

5.3.1 *ACANTHUS EBRACTEATUS* VAHL *(ACANTHACEAE)*

(*Symb. Bot. 2: 75–1791.*)
RLC & C: LC (ver 3.1); YP: 2010; DA: 2008–03–07
E. – Sea Holly; VN: S. Harihusa; H & B. – Harcuch Kanta; Bo. – Nivagur; Tam. – Kalimulli; Tel. – Etichilla.

Diagnostic description (Figures 5.1 and 5.2; Tables 5.1 and 5.3)

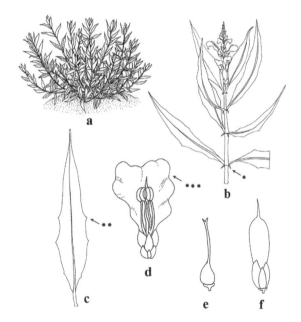

FIGURE 5.1 (a) Canopy; (b) flowering twig (×1/5); (c) leaf (×1/4); (d) flower (×1/2); (e) carpel (×1/1.5); (f) fruit (×1/1.5).

Bushy shrub with dense growth; aerial roots develop occasionally from lower surface of reclining stem; leaves simple, petiolate, decussate; lamina oblong-elliptic, spiny margin, apex mucronate; sometimes tender leaves with entire margin; inflorescence raceme; flower appears to be showy; calyx: sepals 4/5, outer one conspicuous enclosing others; corolla: petal 1, white, zygomorphic; androecium: stamens 4, anthers adhere loosely; gynoecium: carpels 2, syncarpous; fruit capsule; germination incipient vivipary.

FIGURE 5.2 **(See color insert.)** (a) Canopy; (b) a flowering twig; (c) fruits.

TABLE 5.1 Salient Morphometry of *Acanthus ebracteatus*

Habit/organ	Features	Size (range)
Lifeform and canopy	Shrub, thicket bush	1.5–2.0 m
Leaf (margin)	Spiny or entire occasionally	7.0–12.0 cm × 3.5–5.0 cm
Inflorescence	Pedunculate	10.0–14.0 cm
Flower (form)	Zygomorphic	3.0–3.5 cm
	Sepals, variable shape	1.2–1.5 cm
	Petal with three lobes	2.0–2.5 cm × 1.6–2.0 cm
Fruit (type)	Capsule	1.5–2.0 cm

Spot Identification

*Leaf: A pair of spines found at the insertion of each leaf.
**Leaf margin: spiny.
***Flower: showy, petal white or rarely faded violet.

5.3.2 *ACANTHUS ILICIFOLIUS* L. *(ACANTHACEAE)*

(Sp. Pl. 639–1753.)
RLC&C: LC (ver 3.1); YP: 2013; DA: 2010–02–21
E. – Holy Mangrove; VN:S. Harihusa; H & B. – Harcuch Kanta; Bo. –
Nivagur; Tam. – Kalimulli; Tel. – Etichilla

Diagnostic description (Figures 5.3 and 5.4; Tables 5.2 and 5.3)

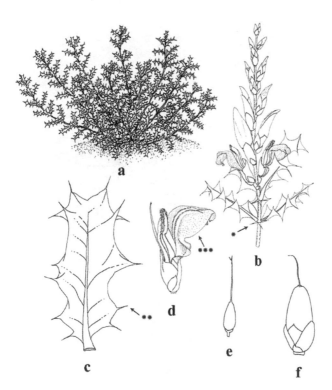

FIGURE 5.3 (a) Canopy; (b) flowering twig (×1/5); (c) leaf (×1/3); (d) flower (×1/2.5);
(e) carpel (×1/1.5); (f) fruit (×1/2).

FIGURE 5.4   **(See color insert.)** (a) Canopy; (b) a flowering twig; (c) fruits.

Bushy shrub with dense growth; aerial roots develop occasionally from lower surface of reclining stem; leaves simple, petiolate, decussate; lamina oblong-elliptic, spiny margin, apex mucronate; sometimes tender leaves with entire margin; inflorescence raceme; flower appears to be showy; calyx: sepals 4/5, outer one conspicuous enclosing others; corolla: petal 1, blue/violet/white occasionally, zygomorphic; androecium: stamens 4, anthers adhere loosely; gynoecium: carpels 2, syncarpous; fruit capsule; germination incipient vivipary.

TABLE 5.2 Salient Morphometry of *A. ilicifolius*

Habit/organ	Features	Size (range)
Lifeform and canopy	Shrub, thicket bush	2.0–2.5 m
Leaf (margin)	Spiny or entire occasionally	6.0–15.0 cm × 2.0–6.0 cm
Inflorescence	Pedunculate	8.0–16.0 cm
Flower (form)	Zygomorphic	4.0–5.5 cm× 3.0–3.8 cm
	Sepals, variable shape	1.7–2.0 cm
	Petal with three lobes	3.5–4.0 cm × 3.0–3.6 cm
Fruit (type)	Capsule	2.0–2.5 cm

Spot Identification

*Leaf: A pair of spines found at the insertion of each leaf.
**Leaf margin: spiny.
***Flower: showy, petal light blue or violet, rarely white.

TABLE 5.3 Comparative Features among Three Species of *Acanthus*

Habit/organ	Attributes	Three species of Acanthus		
		A. ebracteatus	A. ilicifolius	A. volubilis*
Habit	Canopy	○ Densely bushy	○ Densely bushy	● Twining
Leaf	Margin	○ Spiny	○ Spiny	● entire
Flower	Corolla	● white /violet	● blue / violet	● white
	Length	● 2.5 cm	● 5.5 cm	○ 3.0 cm
Stigma	Shape	○ Furcated	○ Furcated	● Obtuse
Fruit	Length (cm)	● 1.5–2.0	● 2.5	● 2.0

● denoting salient feature, ○ denoting common feature; (*A. volubilis- mentioned in 5.5.1 description).

5.3.3 *AEGIALITIS ROTUNDIFOLIA* ROXB. *(PLUMBAGINACEAE/AGIALITIDACEAE)*

(Fl. Ind. 2: 111–1824.)
RLC&C: NT (ver 3.1); YP: 2010; DA: 2008–03–07
VN:B. – Tara

Diagnostic description (Figures 5.5 and 5.6; Table 5.4)

Shrub; trunk base swollen due to fusion of basal roots; fluted axis; leaves simple, alternate, petiole dilated at base encircling the stem; lamina broadly ovate; inflorescence raceme; flower pentamerous; calyx: sepals 5; corolla: petals 5, white, short adjoining at base; androecium: stamens 5, filament base adjoining to form hollow tube; gynoecium: carpels 5, syncarpous; fruit capsule, yellowish at young stage but become brown quickly after maturation; germination incipient vivipary.

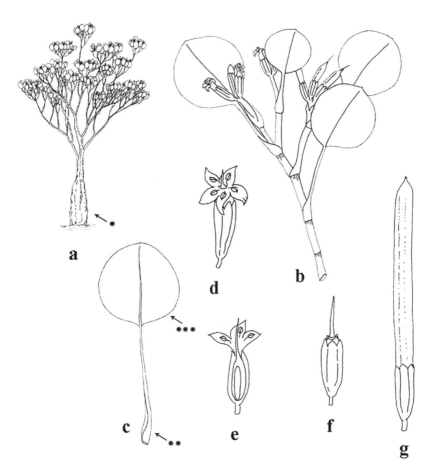

FIGURE 5.5   (a) Canopy; (b) flowering twig (×1/5); (c) leaf (×1/4.5); (d) flower (×1/2); (e) section of flower (×1/2); (f) carpel (×1/1.5); (g) fruit (×1/2).

FIGURE 5.6 **(See color insert.)** (a) Canopy; (b) leaf; (c) flower; (d) a twig with fruits.

TABLE 5.4 Salient Morphometry of *Aegialitis rotundifolia*

Habit/organ	Attributes	Size (range)
Lifeform and canopy	Shrub & under canopy	3.0–4.0 m
Stem base	Pyramid like swollen from base	20.0–40 cm
Leaf	Petiole, sheathing base	5.0–8.5 cm
	Lamina, Broadly ovate	6.0–9.0 cm × 5.0–8.0 cm
Inflorescence	Pedunculate	6.0 cm
Flower (Form)	Pentamerous, Actinomorphic	1.0–1.5 cm × 0.8–1.0 cm
	Sepal, Polypetalous	0.8–1.0 cm × 0.17–0.2 cm
	Petal, mid-vain conspicuous	0.8–1.2 cm × 1.2–0.15 cm
Fruit (Type)	Capsule	6.0–7.0 cm

Spot Identification
*Trunk base: swollen like pyramid from base.
** Petiole: long with sheathing base encircling stem.
*** Lamina: broadly ovate.

5.3.4   *AEGICERAS CORNICULATUM* (L.) BLANCO (*PRIMULACEAE/MYRSINACEAE*)

(Fl. Filip. 79–1837.)
RLC&C: LC (ver 3.1); YP: 2010; DA: 2008–03–07
VN: B. – Khalsi; H.-Halsi; Tam. – Nari-kandam; Tel. – Dudumara

Diagnostic description (Figure 5.7 and 5.8; Table 5.5)

Shrub; bark smooth, pale dark gray; leaves simple, alternate, petiole short; lamina elliptic-obovate; inflorescence umbel at shoot apex; flower pentamerous; calyx: sepals 5, polysepalous; corolla: petals 5, white, twisted; androecium: stamens 5, united at base; gynoecium: ovary one chambered; fruit capsule, terminally curved with pointed apex; germination incipient vivipary.

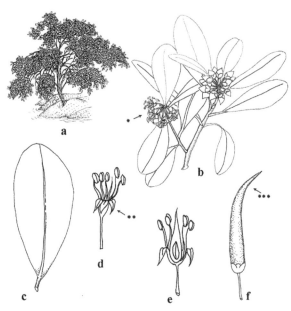

FIGURE 5.7   (a) Canopy; (b) flowering twig (×1/5); (c) leaf (×1/3); (d) flower (×1/2); (e) section of flower (×1/1.5); (f) fruit (×1/3).

FIGURE 5.8 **(See color insert.)** (a) Canopy; (b) flowering buds; (c) flowers; (d) fruits.

TABLE 5.5 Salient Morphometry of *Aegiceras corniculatum*

Habit/organ	Attributes	Size (range)
Lifeform and canopy	Shrub, non-conspicuous canopy	3.5–4.0 m
Leaf	Petiole, short	0.3–0.5 cm
	Lamina, elliptic-obovate	6–10 cm × 3–5 cm
Flower (Form)	Showy	1.5–2.0 cm× 0.5–1.0 cm
Petal	White, twisted	0.8–1.5×0.24–0.26 cm
Fruit (Type)	Capsule	5.0–7.5 cm

Spot Identification
*Inflorescence: umbel at shoot apex.
**Flower: white, showy with strong odor after full bloom.
***Fruit: terminally curved with pointed apex.

5.3.5 *AVICENNIA ALBA* BLUME *(ACANTHACEAE/ AVICENNIACEAE)*

(Bijdr. 821–1826.):* considered as a synonym of ***Avicennia marina*** (Forssk.) Vierh. as per all three webs (www.theplantlist.org, www.catalogoflife.org and www.gbif.org), but the fact is supposed not to be true as the authors verified the attributes of *A. alba* in the field; Peter Saenger also agrees with the authors' view while reviewing the draft.
RLC&C: LC (ver 3.1); YP: 2010; DA: 2008–03–07
VN: B. – Kalo ban.

Diagnostic description (Figures 5.9 and 5.10; Tables 5.6 and 5.9)

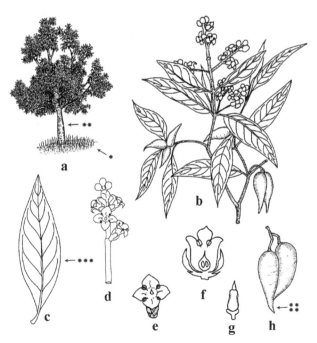

FIGURE 5.9   (a) Canopy; (b) flowering twig (×1/5); (c) leaf (×1/4); (d) inflorescence (×1/3); (e) flower (×1/1.5); (f) section of flower (×1/1.5); (g) carpel (×1); (h) fruit (×1/2).

FIGURE 5.10 **(See color insert.)** (a) Canopy; (b) leaf; (c) flowering twig; (d) fruits.

Medium tree; bark grayish black; aerial roots exhibit pencil like upward projection known as breathing roots (pneumatophores); leaves simple, petiolate, decussate; lamina lanceolate, acuminate, silvery gray & pubescent at ventral side; inflorescence spicate raceme; flowers sessile; calyx: sepals 5, polysepalous; corolla: petals 4, gamopetalous, orange-yellow;

androecium: stamens 4, epipetalous; gynoecium: ovary syncarpous; fruit exhibiting conical shape with terminal beak; germination incipient vivipary.

TABLE 5.6   Salient Morphometry of *Avicennia alba*

Habit/organ	Attributes	Size (range)
Lifeform and canopy	Tree	15–20 m
Aerial roots	Breathing root	10.0–30.0 cm
Leaf	Petiole, pulvinus	1.5–2.0 cm
	Lamina, lanceolate	8.0–12.5 cm × 1.5–4.5 cm
Inflorescence (Form)	Spicate raceme	5.0–10 cm
Flower (Form)	Small, sessile	0.5–0.8 cm × 0.25–0.35 cm
Fruit (Type)	Capsule, pubescent	2.0–4.0 cm

Spot Identification
*Trunk base: surrounded with dense aerial roots in the ground.
**Bark: grayish black.
***Leaf: lanceolate, acuminate.
****Fruit: conical with terminal beak.

5.3.6 *AVICENNIA MARINA* (FORSSK.) VIERH. (ACANTHACEAE/AVICENNIACEAE)

(Denkschr. Kaiserl. Akad. Wiss., Wien. Math.-Naturwiss. Kl. 71: 435–1907.)
RLC&C: LC (ver 3.1); YP: 2010; DA: 2008–03–07
E. – Gray mangrove; VN:B. – Peyara ban

Diagnostic description (Figures 5.11 and 5.12; Tables 5.7 and 5.9)

Shrub to tree; bark smooth, yellowish brown, not fissured; aerial roots exhibit pencil like upward projection known as breathing roots (pneumatophores); leaves simple, petiolate, decussate; lamina elliptic-oblong, acute, whitish & pubescent at ventral side; inflorescence compound spike; flowers sessile; calyx: sepals 5, polysepalous; corolla: petals 4, gamopetalous, pale or orange-yellow; androecium: stamens 4, epipetalous; gynoecium: ovary syncarpous; fruit capsule, spherical, shortly beaked; germination incipient vivipary.

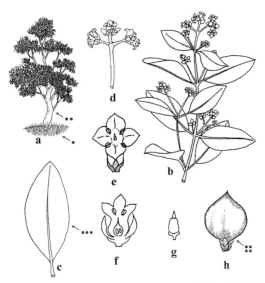

FIGURE 5.11   (a) Canopy; (b) flowering twig (×1/5); (c) leaf (×1/4); (d) inflorescence (×1/3); (e) flower (×1); (f) section of flower (×1); (g) carpel (×1); (h) fruit (×1/2).

FIGURE 5.12   **(See color insert.)** (a) Canopy; (b) leaf; (c) flowering twig; (d) fruit.

TABLE 5.7 Salient Morphometry of *A. marina*

Habit/organ	Features	Size (range)
Lifeform and canopy	Shrub, small to medium tree	3.0–15.0 m
Aerial roots	Breathing root	10.0–50 cm
Leaf	Petiole, pulvinus	0.5–1.5 cm
	Lamina, elliptic-oblong	5.0–14.0 cm × 1.5–3.5 cm
Inflorescence (Form)	Compound spike	5.0–10.0 cm
Flower (Form)	Small, sessile	0.5–0.9 cm × 0.3–0.5 cm
Fruit (Type)	Capsule, shortly beaked	1.2–2.5 cm

Spot Identification

*Trunk base: surrounded with dense aerial roots in the ground.

**Bark: yellowish brown, not fissured.

***Leaf: elliptic-oblong, acute.

****Fruit: spherical, shortly beaked.

5.3.7 AVICENNIA OFFICINALIS L. (ACANTHACEAE/ AVICENNIACEAE)

(Sp. Pl. 110–1753.)
RLC&C: LC (ver 3.1); YP: 2010; DA: 2008–03–07
VN:B. – Jat ban; H.-Bina; Bo. – Tivar; S. – Tuvara; Tel. – Nallamanda;
Tam. – Madaipattai

Diagnostic description (Figures 5.13 and 5.14; Tables 5.8 and 5.9)

Large tree; bark smooth, dark gray; aerial roots exhibit pencil like upward projection known as breathing roots (pneumatophores); leaves simple, petiolate, decussate; lamina broadly ovate-oblong, obtuse; inflorescence compound spike, flowers 10–12, sessile; calyx: sepals 5, polysepalous; corolla: petals 4, gamopetalous, orange-yellow; androecium: stamens 4, epipetalous; gynoecium: ovary syncarpous; fruit capsule, flattened spherical; germination incipient vivipary.

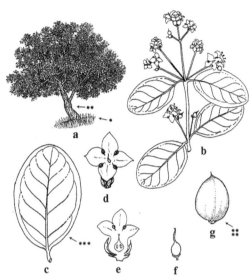

FIGURE 5.13 (a) Canopy; (b) flowering twig (×1/5); (c) leaf (×1/4); (d) flower (×1/1.5); (e) section of flower (×1/1.5); (f) carpel (×1/1.5); (g) fruit (×1/2.5).

FIGURE 5.14 **(See color insert.)** (a) Canopy; (b) leaf; (c) flowering twig; (d) fruits.

TABLE 5.8   Salient Morphometry of *A. officinalis*

Habit/organ	Features	Size (range)
Lifeform and canopy	Tree with wide girth	15–25 m
Aerial roots	Breathing root	10.0–30.0 cm
Leaf	Petiole, pulvinus	1.2–1.5 cm
	Lamina	8.0–12.0 cm × 4.5–5.5 cm
Inflorescence	Compound spike	10.0–25.0 cm
Flower (Form)	Medium, sessile	0.5–1.0 cm × 1.0–1.5 cm
Fruit (Type)	Capsule, flattened spherical	0.28–3.0 cm × 2.5–2.8 cm

Spot Identification

*Trunk base: surrounded with dense aerial roots in the ground.

**Bark: dark gray.

***Leaf: broadly ovate-oblong, obtuse.

****Fruit: flattened spherical.

TABLE 5.9   Comparative Features among Three Species of *Avicennia*

Habit/organ	Attributes	Three species of Avicennia		
		A. alba	**A. marina**	**A. officinalis**
Bark	Color	● grayish black	● yellowish brown	● dark gray
Leaf	Apex	● acuminate	● acute	● obtuse/blunt
Inflorescence	Type	● spicate raceme	○ compound spike	○ compound spike
Fruit	Shape	● conical with beak	○ spherical	○ flattened spherical
	Length	○ 2.0–4.0 cm	○ 2.5 cm	○ 3.0 cm × 2.8 cm

● denoting salient feature, ○ denoting common feature.

5.3.8 *BRUGUIERA CYLINDRICA* (L.) BLUME (RHIZOPHORACEAE)

(Enum. Pl. Javae 1: 91–1827.)

RLC&C: LC (ver 3.1); YP: 2010; DA: 2008–03–07

VN:B. – Bakul kakra

Diagnostic description (Figures 5.15 and 5.16; Tables 5.10 and 5.14)

Medium to tall tree; bark greenish; underground horizontal roots occasionally exhibit knee like outer projection; leaves simple, petiolate, decussate; lamina ovate-lanceolate, acute; inflorescence cyme, pedicellate, 3 flowers at each peduncle; flower small; calyx: sepals 8, whitish green; corolla: petals 8; androecium: stamens 16, free but eight groups; gynoecium: carpels 2–3, syncarpous; germination vivipary; hypocotyls, each with blunt end, hanging down from branches.

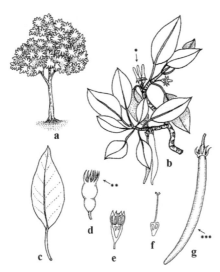

FIGURE 5.15 (a) Canopy; (b) flowering twig (×1/5); (c) leaf (×1/4); (d) flower (×1/2); (e) section of flower (×1/1.5); (f) carpel (×1/1.5); (g) hypocotyl (×1/4).

FIGURE 5.16 **(See color insert.)** (a) Canopy; (b) flowering twig; (c) hypocotyls hanging from mother tree; (d) hypocotyl.

TABLE 5.10   Salient Morphometry of *Bruguiera cylindrica*

Habit/organ	Features	Size (range)
Lifeform and canopy	Tree	11.0–15.0 m
Lamina (shape)	Ovate-lanceolate	10.0–12.0 cm × 4.0–5.0 cm
Inflorescence	Cyme	2.5–3.0 cm
Flower	Small	1.0–1.3 cm × 0.35–0.5 cm
Calyx	Whitish green	0.35–0.5 cm
Hypocotyl (Length)	Round rod	14.0–16.0 cm

Spot Identification
* Inflorescence: 3 flowers in each peduncle.
** Calyx: sepals 8, whitish green.
*** Hypocotyl: 14.0–16.0 cm long.

5.3.9   *BRUGUIERA GYMNORHIZA* (L.) SAVIGNY (RHIZOPHORACEAE)

(Encycl. 4: 696–1798.); Bruguiera gymnorhiza (L.) Lam. Synonym of *Bruguiera gymnorhiza* (L.) Savigny
RLC&C: LC (ver 3.1); YP: 2010; DA: 2008–03–07
E. – Oriental mangrove; VN:B. – Kakra

[Note: The name of species, *Bruguiera gymnorhiza* first appeared in the book, *Encyclopédie méthodique, Botanique* edited by Lamarck and Poiret (1798), which was based on *Rhizophora gymnorhiza* L. Originally, J.C. Savigny (1777–1851) described this species in his publication edited by Lamarck and Poiret (1798) that has always been cited as *Bruguiera gymnorhiza* (L.) Lamarck. However, the author citation of this species should be as *Bruguiera gymnorhiza* (L.) Savigny ex Lam. & Poiret 1798 (Le palatuvier; Bruguiera Lam. Encyclopédie Méthodique, Botanique *4*(2), 696). But for most purposes *B. gymnorhiza* (L.) Savigny is cited. Later, Stafleu and Cowan (1985) also mentions *B. gymnorhiza* (L.) Savigny to be the correct author citation instead of *B. gymnorhiza* (L.) Lam.]

Diagnostic description (Figures 5.17 and 5.18; Tables 5.11 and 5.14)

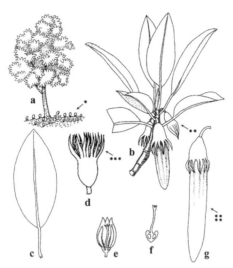

FIGURE 5.17 (a) Canopy; (b) flowering twig (×1/5); (c) leaf (×1/5); (d) flower (×1/2); (e) section of flower (×1/2.5); (f) section of carpel (×1/1.5); (g) hypocotyl (×1/4).

FIGURE 5.18 (See color insert.) (a) Canopy; (b) knee roots; (c) young tree; (d) hypocotyls hanging from mother tree.

Medium to tall tree; bark reddish/black green; aerial knee roots, buttress at trunk base; leaves simple, petiolate, decussate; lamina ovate lanceolate, reddish green, acute; inflorescence solitary, cyme; flower showy, pedicellate; calyx: sepals 13–16, reddish; corolla: petals 13–16, white, 3–4 terminal cilia; androecium: stamens 26–32, free; gynoecium: carpels 3, syncarpous; germination vivipary; hypocotyls, each with reddish sepal lobes and round end, hanging down from branches.

TABLE 5.11　Salient Morphometry of *B. gymnorhiza*

Habit/organ	Features	Size (range)
Lifeform and canopy	Tree	12.0–15.0 m
Lamina (shape)	Ovate-lanceolate	13.0–21.0 cm × 5.0–6.5 cm
Inflorescence	Solitary, cyme	
Flower	Large	4.5–5.5 cm × 0.4–0.6 cm
Calyx	Reddish	1.0–1.5 cm
Hypocotyle (Length)	Cylindrical	16.0–18.0 cm

Spot Identification
* Aerial root: knee root and root buttress.
** Inflorescence: solitary.
*** Calyx: 13–16, reddish.
**** Hypocotyl: 18 cm long with scarlet red calyx lobes.

5.3.10　*BRUGUIERA PARVIFLORA* (ROXB.) WIGHT & ARN. EX GRIFF. *(RHIZOPHORACEAE)*

(Trans. Med. Soc. Calcutta 8: 10–1836.)
RLC&C: LC (ver 3.1); YP: 2010; DA: 2008–03–07
E. – Small flower mangrove; VN:B. – Bakul kakra

Diagnostic description (Figures 5.19 and 5.20; Tables 5.12 and 5.14)

Medium to tall tree; bark blackish gray, fissured; roots buttress and knee roots; leaves simple, petiolate, decussate; lamina ovate-lanceolate, acute; inflorescence cyme, 4–7 flowers at each peduncle; flower small; calyx: sepals 8–10, white greenish; corolla: petals 8; androecium: stamens 16, free; gynoecium: carpels 3, syncarpous; germination vivipary; hypocotyls, each with long stripes and pointed end, hanging down from branches.

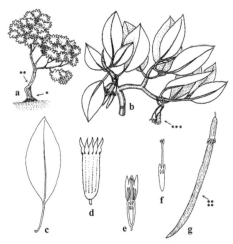

FIGURE 5.19   (a) Canopy; (b) flowering twig (×1/4); (c) leaf (×1/3); (d) flower (× 0.5); (e) section of flower (×0.5); (f) section of carpel (×0.5); (g) hypocotyl (×1/4.5).

FIGURE 5.20   (See color insert.) (a) Canopy;*b*. knee roots; (c) flowering twig; (d) hypocotyls hanging from mother tree.

TABLE 5.12   Salient Morphometry of *B. parviflora*

Habit/organ	Features	Size (range)
Lifeform and canopy	Tree	12.0–15.0m
Lamina (shape)	Ovate-lanceolate, acute	10.0–14.0× 4.0–5.5 cm
Inflorescence	4–7 flowers at each peduncle	
Flower	Small	1.0–1.2 cm × 0.35–0.5 cm
Calyx	Greenish	0.15–0.25 cm
Hypocotyl (Length)	Round rod	20.0–25.0 cm

Spot Identification

*Aerial root: knee root and root buttress.

**Bark: blackish gray, fissured.

***Inflorescence: 4–7 flowers at each peduncle.

****Hypocotyl: up to 25 cm long with long stripes and pointed end.

5.3.11   *BRUGUIERA SEXANGULA* (LOUR.) POIR. (*RHIZOPHORACEAE*)

(Encycl. Suppl. 4: 262–1816.)
RLC&C: LC (ver 3.1); YP: 2010; DA: 2008–03–07
VN:B. – Kakra

Diagnostic description (Figures 5.21 and 5.22; Tables 5.13 and 5.14)

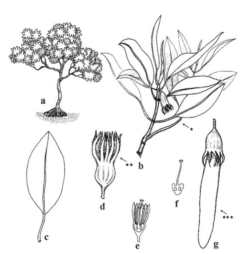

FIGURE 5.21   (a) Canopy; (b) flowering twig (×1/4); (c) leaf (×1/3); (d) flower (×1/2); (e) section of flower (×1/1.5); (f) section of carpel (×1/1.5); (g) hypocotyl (×1/4).

FIGURE 5.22   (See color insert.) (a) Canopy; (b) flowering twig; (c) hypocotyl.

Small to medium tree; bark yellowish green; root buttress; leaves simple, petiolate, decussate; lamina ovate-lanceolate, acute; inflorescence solitary, cyme; flower medium; calyx: sepals 10–12, yellowish; corolla: petals 10–12; androecium: stamens 20–24, free; gynoecium: carpels 3, syncarpous; germination vivipary; hypocotyls, each with yellowish sepal lobes and round end, hanging down from branches.

TABLE 5.13   Salient Morphometry of *B. sexangula*

Habit/organ	Features	Size (range)
Lifeform and canopy	Tree	9.0–12.0 m
Lamina (shape)	Ovate-lanceolate	9.5–13.0 cm × 3.3–6.5 cm
Inflorescence	Solitary, cyme	
Flower	Medium	2–3 cm × 0.5 cm
Calyx	Yellowish	0.75–1.0 cm
Hypocotyl (Length)	Cylindrical	8.0–12.0 cm

Spot Identification
* Inflorescence: solitary.
** Calyx: sepals 10–12, yellowish.
*** Hypocotyl: 8–12 cm long with yellowish sepal lobes and round end.

TABLE 5.14   Comparative Features among Four Species of *Bruguiera*

Habit/organ	Attributes	Four species of Bruguiera			
		B. cylindrica	**B. gymnorhiza**	**B. parviflora**	**B. sexangula**
Bark	Color	○ Greenish	○ Reddish-black	○ Blackish gray	○ Yellowish green
Inflorescence	Type	● 3 flowers	○ Solitary	● 4–7 flowers	○ Solitary
Sepals	Numbers	● 8	● 13–16	● 8–10	● 10–12
	Color	○ Greenish	● Reddish	○ White-greenish	● Yellowish
Hypocotyl	length	●14.0–16.0 cm	●16–18 cm	●20.0–25.0 cm	●8–12 cm

● denoting salient feature, ○ denoting common feature.

5.3.12   *CERIOPS DECANDRA* (GRIFF.) DING HOU / **CERIOPS DECANDRA* (GRIFF.) W. THEOB. (RHIZOPHORACEAE)

(Ding Hou: Hou, Flora Malesiana, Ser.1. Vol.5:471.1958. propart.: Burmah ed. 3, 2: 480–1860.): * It is a synonym of *Ceriops decandra* (Griff.) Ding Hou (explanation below).
RLC&C: NT (ver 3.1); YP: 2010; DA: 2008–03–07
VN:B. – Jale-Goran or jhamti-goran

[Note: The two accepted names of *Ceriops decandra* are ambiguous as well as conflicting in term of author citation, as provided by two different webs, www.theplantlist.org and http://www.catalogoflife.org. The former one mentions that *C. decandra* (Griff.) W. Theobald (1883) as an accepted name. On the contrary, the web, http://www.catalogoflife.org treats this name, *C. decandra* (Griff.) W. Theobald as an ambiguous synonym of *C. decandra* (Griffith) Ding Hou (1958).

The name of this species was adopted by Roxburgh in his Hortus Bengalensis (1814) as *Rhizophora decandra* without adding any description or diagnosis. Later, Griffith (1836 & 1854) validated this name by providing description. Griffith (1836) proposed a new combination,

Bruguiera decandra (Roxb.) W. Griffith, based on *Rhizophora decandra* Roxb. However, in his later publication Griffith (1854) retained the earlier name.

Mabberley (1985) stated that Theobald's binomial was based on *Rhizophora decandra* Roxb. ex Griff. (1854) whereas Ding Hou's binomial was based on *Bruguiera decandra* (Roxb.) Griff. (1836). Both Theobald and Ding Hou treated this species under the same genus *Ceriops*. Apparently Ding Hou's new combination is a later homonym, but it was based on earlier legitimate name, e.g., *Bruguiera decandra* (Roxb.) Griff. (1836). The Article 11.4 of ICN (Melbourne Code, 2011) stated that "For any taxon below the rank of genus, the correct name is the combination of the final epithet of the earliest legitimate name of the taxon ...." Therefore, *Ceriops decandra* (Griffith) Ding Hou (1958) is the correct name of this species. However, Sheue et al., (2009) seperated *Ceriops decandra* (Griffith) Ding Hou into three species such as *Ceriops decandra, C. pseudodecandra* and *Ceriops zippeliana.*]

Diagnostic description (Figures 5.23 and 5.24; Tables 5.15 and 5.17)

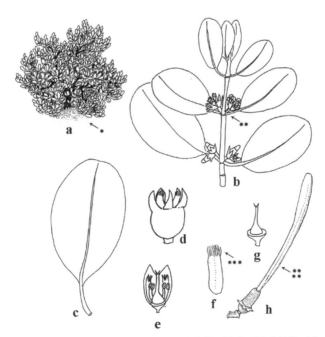

FIGURE 5.23   (a) Canopy; (b) flowering twig (×1/5); (c) leaf (×1/2.5); (d) flower (×1); (e) section of flower (×1/1.5); (f) petal (×0.75); (g) carpel (×1); (h) hypocotyl (×1/2).

FIGURE 5.24 **(See color insert.)** (a) Canopy; (b) flowering twig; (c) hypocotyls attached with mother tree.

Bushy shrub; bark pale yellow; broom like stilt roots at the trunk base; leaves simple, petiolate, decussate; lamina ovate-elliptic, obtuse or round apex; inflorescence umbel-like cyme; flowers buds 16 or more at single peduncle, but 4–6 get mature; calyx: sepals 5, yellowish green; corolla: petals 5, each petal ended with ciliated appendages; androecium: stamens 10, free; gynoecium: carpels 3, syncarpous; germination vivipary; hypocotyls, each is usually found projected upward with long stripes and blunt end, hanging down from branches.

TABLE 5.15   Salient Morphometry of *Ceriops decandra*

Habit/organ	Features	Size (range)
Lifeform and canopy	Bushy shrub, under canopy	3.0–5.0 m
Lamina (shape)	Ovate-elliptic	9.0–10.0 cm × 4.5–5.5 cm
Inflorescence	Umbel-like cyme	1.0–2.0 cm
Flower	Small, 16 numbers	0.75–1.0 cm × 0.4–0.5 cm
Calyx	Yellowish green	0.3–0.4 cm ×0.1–0.15 cm
Hypocotyl (Length)	Round rod	8.0–12.0 cm

Spot Identification

*Bushy shrub: canopy appears sprawling.

**Inflorescence: 16 buds projected upward at single peduncle.

***Petal: each petal ended with ciliated appendages.

****Hypocotyl: 8–12 cm long hypocotyl projected upward with long stripes and blunt end.

5.3.13   *CERIOPS TAGAL* (PERR.) C.B. ROB. *(RHIZOPHORACEAE)*

(Philipp. J. Sci., C 3: 306–1908.)
RLC&C: LC (ver 3.1); YP: 2010; DA: 2008–03–07
VN: *Mat-Goran or jat-goran;* Tam. – Pandikutti

Diagnostic description (Figures 5.25 and 5.26; Tables 5.16 and 5.17)

Small tree; pyramidal look; broom like stilt roots at the trunk base; leaves simple, petiolate, decussate; lamina elliptic, obtuse; inflorescence cyme, arising from terminal shoot and bending downward; flower buds 10–12 at single peduncle; calyx: sepals 5, greenish; corolla: petals 5, each petal ended with three small glands like appendages; androecium: stamens 10, free; gynoecium: carpels 3, syncarpous; germination vivipary, hypocotyl, each with long stripes and pointed end, hanging down from branches.

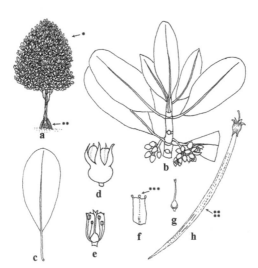

FIGURE 5.25   (a) Canopy; (b) flowering twig (×1/5); (c) leaf (×1/2.5); (d) flower (×1); (e) section of flower (×1/1.5); (f) petal (×0.75); (g) carpel (×1); (h) hypocotyl (×1/4.5).

FIGURE 5.26 **(See color insert.)** (a) Canopy; (b) flowering twig; (c) hypocotyls hanging from mother tree.

TABLE 5.16   Salient Morphometry of *C. tagal*

Habit/organ	Features	Size (range)
Lifeform and canopy	Small tree with pyramidal canopy	5.0–6.5 m
Lamina (shape)	Elliptic	8.0–9.4 cm × 4.0–4.5 cm
Inflorescence	Downward cyme	1.5–2.0 cm
Flower	Small, 10–12 numbers	0.8–1.2 cm × 0.3–0.4 cm
Calyx	Yellowish	0.4–0.45 cm ×0.2–0.25 cm
Hypocotyl (Length)	Round rod	20.0–25.0 cm

Spot Identification

*Small tree: canopy looks pyramid.

**Aerial root: broom like stilt roots at the trunk base.

***Petal: each petal ended with three glands like appendages.

****Hypocotyl: 20–25 cm long, hanging down, with long stripes and pointed end.

TABLE 5.17   Comparative Features among Two Species of *Ceriops*

Habit/organ	Attributes	Two species of Ceriops	
		C. decandra	**C. tagal**
Canopy	Shape	• Bush appears sprawling	• Small tree looks pyramid
Inflorescence	Projection	• Upward	• Downward
Flowers bud	Numbers	○16	○10–12
Petal	Appendage	• Ciliated terminally	•Threads with small head
Hypocotyl	Length	• 8–12 cm	• 20.0–25.0 cm

• denoting salient feature, ○ denoting common feature.

5.3.14   *EXCOECARIA AGALLOCHA L. (EUPHORBIACEAE)*

(Syst. Nat. ed. 10–2: 1288–1759.)
RLC&C: LC (ver 3.1); YP: 2010; DA: 2008–03–07
VN: Gneoa; Bo. – Geva; S. – Agaru; Tel. – Thilla; Tam. – Kampetti

Diagnostic description (Figures 5.27 and 5.28; Table 5.18)

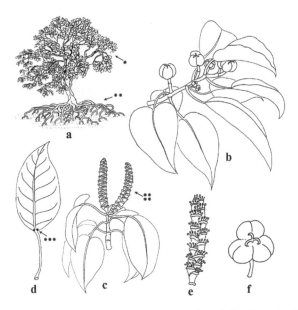

FIGURE 5.27   (a) Canopy; (b) female flowering twig (×1/5); (c) male flowering twig (×1/5); (d) leaf (×1/2); (e) male inflorescence (×1/2); (f) female flower (×1/1.5).

FIGURE 5.28 **(See color insert.)** (a) Canopy of male tree; (b) root; (c) male inflorescence; (d) female inflorescence.

Medium tree, dioecious; bark fissured; branch and leaves with milky latex; aerial roots spreading horizontal like serpentine appearance; leaves simple, petiolate with a pair of glands at lamina base, alternate, become reddish at maturity; male inflorescence catkin, yellow, male flowers sessile; perianth: tepals 3, polytepalous; androecium: stamens 3, free; female inflorescence mixed cyme, female flower sessile; perianth: tepals 3, polyte-palous; gynoecium: carpels 3 with distinct locules; germination epigeal.

TABLE 5.18   Salient Morphometry of *Excoecaria agallocha*

Habit/organ	Features	Size (range)
Lifeform and canopy	Medium tree	15.0–20.0 m
Lamina (shape)	Ovate-elliptic	4.0–7.5 cm × 2.4–3.5 cm
Inflorescence (Male)	Catkin	4.0–7.0 cm
Inflorescence (Female)	Ovary with three locules	1.0–1.5 cm

Spot Identification

*Milky latex: bark, branch, leaves.

**Aerial root: spreading horizontal like serpentine appearance.

*** Leaf: petiolate with a pair of glands at lamina base.

****Inflorescence: male inflorescence catkin with yellow color.

5.3.15   *HERITIERA FOMES* BUCH.-HAM. *(MALVACEAE/ STERCULIACEAE)*

(Embassy ed. 2, 3: 319–1800.)

RLC&C: EN (ver 3.1); YP: 2010; DA: 2008–03–07

VN: B. – Sundari

Diagnostic description (Figures 5.29 and 5.30; Tables 5.19 and 5.21)

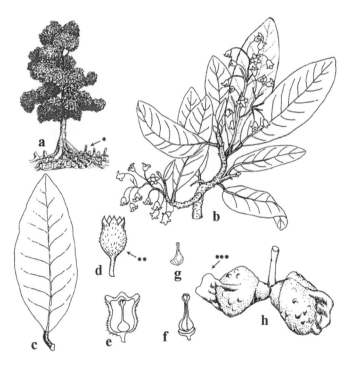

FIGURE 5.29   (a) Canopy; (b) flowering twig (×1/5); (c) leaf (×1/4); (d) flower (×1/1.5); (e) section of flower (×1/1.5); (f) carpel (×1/1.5); (g) anther (×1/1.5); (h) fruit (×1/2).

FIGURE 5.30   **(See color insert.)** (a) Flowering twigs; (b) leaf; (c) fruiting branch.

Medium to tall tree; bark gray brown outer, but inner turn reddish after peeling; aerial roots (pneumatophores) peg-like blunt end and become exposed as plank-like horizontal ribbon with buttresses at trunk base; leaves simple, petiolate, alternate; lamina elliptic-lanceolate, coriaceous; inflorescence mixed cyme; flowers unisexual, pedicellate; calyx: sepals 4–5, bell-shaped with pink eye inside, gamosepalous; corolla absent; androecium: stamens 5, anthers fused around central column representing as pistillode; gynoecium: carpels 4–5, apocarpous; fruit with keel.

TABLE 5.19   Salient Morphometry of *Heritiera fomes*

Habit/organ	Features	Size (range)
Lifeform and canopy	Tree	10.0–25.0 m
Lamina (shape)	Elliptic-lanceolate	8.0–15.0 cm × 3.0–5.0 cm
Inflorescence	Mixed cyme	4.5–6.0 cm
Flower	Small with bell-shaped	0.7–1.0 cm × 0.3–0.5 cm
Fruit	With keel	1.8–2.0 cm × 4.5–5.0 cm

Spot Identification
*Aerial roots: pneumatophore with peg-like blunt end and buttress at trunk base.
**Flower: bell-shaped with pink eye inside.
***Fruit: with keel.

5.3.16 HERITIERA LITTORALIS AITON (MALVACEAE/ STERCULIACEAE)

(Hort. Kew. 3: 546–1789.)
RLC&C: LC (ver 3.1); YP: 2010; DA: 2008–03–07
VN: Sundari

Diagnostic description (Figures 5.31 and 5.32; Tables 5.20 and 5.21)

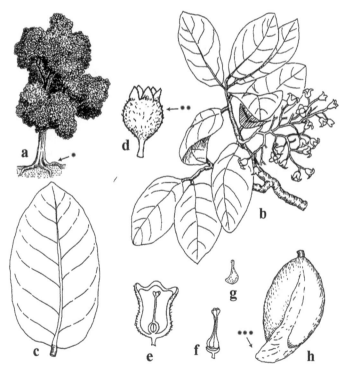

FIGURE 5.31   (a) Canopy; (b) flowering twig (×1/5); (c) leaf (×1/5); (d) flower (×1/1.5); (e) section of flower (×1/1.5); (f) carpel (×1/1.5); (g) anther (×1/1.5); (h) fruit (×1/3).

FIGURE 5.32   (See color insert.) (a) Canopy; (b) inflorescence; (c) a twig with fruits; (d) fruit.

Tall tree; bark dark or pinkish gray; buttresses at the base of the trunk and form plank-like ribbon; leaves simple, petiolate, alternate; lamina oblong-lanceolate, coriaceous; inflorescence mixed cyme; flowers appear in cluster, unisexual, bell-shaped; sepals 4–5, gamosepalous, petals absent; androecium: stamens 5, anthers fused around central column representing as pistillode; gynoecium: carpels 4–5, apocarpous; fruit with large keel.

TABLE 5.20   Salient Morphometry of *H. littoralis*

Habit/organ	Features	Size (range)
Lifeform and canopy	Tree	20.0–35.0 m
Lamina (shape)	Oblong-lanceolate	12.0–18.0 cm × 8.0–12.0 cm
Inflorescence	Mixed cyme	4.0–5.0 cm
Flower	Small with bell-shaped	0.5–0.8 cm × 0.3–0.4 cm
Fruit	With large keel	5.0–6.0 cm × 3.5–4.0 cm

Spot Identification
*Aerial roots: buttresses at the base of the trunk and form plank-like ribbon.
**Flower: bell-shaped.
***Fruit: with 3.5–4.0 cm long flattened keel.

TABLE 5.21 Comparative Features among Two Species of *Heritiera*

Habit/organ	Attributes	Two species of Heritiera	
		H. fomes	**H. littoralis**
Aerial roots	Form	● Pneumatophores	○ Buttress
Leaf	Shape	○ Elliptic-lanceolate	○ Oblong-lanceolate
Fruit	Keel	○ Not distinct	● Large

● denoting salient feature, ○ denoting common feature.

5.3.17 KANDELIA CANDEL (L.) DRUCE (RHIZOPHORACEAE)

(Bot. Exch. Club Soc. Brit. Isles 3: 420–1914.)
RLC&C: LC (ver 3.1); YP: 2010; DA: 2008–03–07
VN: Goria; Tam. – Kandal; Kan. – Kandale; Tel. – Kandigala; Mal. – Kantal

Diagnostic description (Figurew 5.33 and 5.34; Table 5.22)

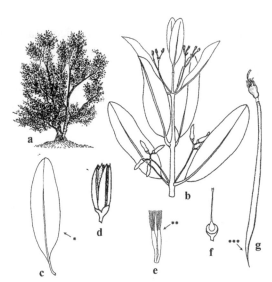

FIGURE 5.33   (a) Canopy; (b) flowering twig (×1/5); (c) leaf (×1/3.5); (d) flower (×1/2);
(e) petal (×1/2); (f) carpel (×1); (g) fruit (×1/7).

FIGURE 5.34   (See color insert.) (a) Canopy; (b) a bunch of hypocotyls hanging from mother tree; (c) hypocotyl.

Shrub to small tree; leaves simple, petiolate, decussate; lamina oblong-lanceolate, acute; inflorescence cyme, 4 fowers at each peduncle; calyx: sepals 5, polysepalous; corolla: petals 5, white, polypetalous, each petal with apical wrinkle filamentous appendages; androecium: stamens 34 or indefinite, free; gynoecium: carpels 3, syncarpous; germination vivipary, hypocotyl, each with smooth surface and long tapering end, hanging down from branches.

TABLE 5.22   Salient morphometry of *Kandelia candel*

Habit/organ	Features	Size (range)
Lifeform and canopy	Small tree	4.0–7.5 m
Lamina (shape)	Oblong-lanceolate	10.0–14.0 cm × 4.0–5.0 cm
Inflorescence	4 flowers	4.5–5.0 cm
Petal	Apical filamentous appendages	0.4–0.5 cm
Hypocotyl (Length)	Round rod with tapering end	40.0–45.0 cm

Spot Identification
*Lamina: oblong-lanceolate.
**Petal: each petal with apical wrinkle filamentous appendages.
***Hypocotyl: 40–45 cm long, hanging down, with smooth surface and long tapering end.

5.3.18 *LUMNITZERA LITTOREA* (JACK) VOIGT (COMBRETACEAE)

(Hort. Suburb. Calcutt. 39–1845.)
RLC&C: LC (ver 3.1); YP: 2010; DA: 2008–03–07
VN: B. – Kripa

Diagnostic description (Figures 5.35 and 5.36; Tables 5.23 and 5.25)

Small tree; bark fissured; leaves simple, petiolate, cyclic; lamina subulate, margin slightly wavy, apical notch; inflorescence terminal raceme; flowers pedicellate; calyx: sepals 5, gamosepalous, forming a tube; corolla: petals 5, polypetalous, dark red, showy; androecium: stamens less than 10, free; gynoecium: ovary 1, stigma sticky.

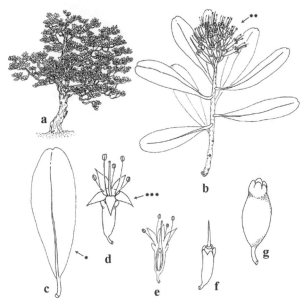

FIGURE 5.35 (a) Canopy; (b) flowering twig (×1/4); (c) leaf (×1/2); (d) flower (×1/1.5); (e) section of flower (×1/1.5); (f) carpel (×1/1.5); (g) fruit (×1/1.5).

FIGURE 5.36 **(See color insert.)** (a) Canopy; (b) flowering twig; (c) fruits.

TABLE 5.23   Salient Morphometry of *Lumnitzera littorea*

Habit/organ	Features	Size (range)
Lifeform and canopy	Small tree	5.0–6.0 m
Lamina (shape)	Subulate	6.0–7.5 cm × 3.0–4.0 cm
Inflorescence	Terminal raceme	2.0–3.0 cm
Flower	Rosaceous	1.2–1.5 cm

Spot Identification
*Lamina: subulate with apical notch.
**Inflorescence: terminal raceme.
***Corolla: petals dark red, showy.

5.3.19 *LUMNITZERA RACEMOSA* WILLD. *(COMBRETACEAE)*

(Neue Schriften Ges. Naturf. Freunde Berlin 4: 187–1803.)
RLC&C: LC (ver 3.1); YP: 2010; DA: 2008–03–07
VN: B. – Kripa; Tel. – Kadivi, Thandara; Tam. – Tipparathai; Mal. – Katak-kantal; Odi. – Tunda

Diagnostic description (Figures 5.37 and 5.38; Tables 5.24 and 5.25)

Shrub; aerial roots appear to be looping from lateral roots, but no aerial root are recorded when species grows in area away from tidal zone; leaves simple, petiolate, cyclic; lamina subulate, margin slightly wavy, apical notch; inflorescence axillary raceme; flowers pedicellate; calyx: sepals 5, gamosepalous, forming a tube; corolla: petals 5, white, polypetalous; androecium: stamens 10, free, arranged in two whorls; gynoecium: ovary 1; stigma absent.

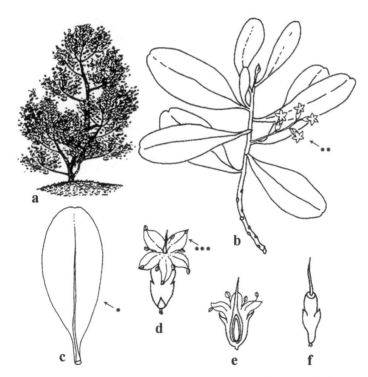

FIGURE 5.37   (a) Canopy; (b) flowering twig (×1/4); (c) leaf (×1/2); (d) flower (×1/1.5); (e) section of flower (×1/1.5); (f) fruit (×1/1.5).

FIGURE 5.38   (See color insert.) (a) Canopy; (b) flowering twig; (c) flowers.

TABLE 5.24   Salient Morphometry of *L. racemosa*

Habit/organ	Features	Size (range)
Lifeform and canopy	Shrub, under canopy	3.0–4.0 m
Lamina (shape)	Subulate	7.0–8.5 cm × 3.0–8.0 cm
Inflorescence	Axillary raceme	6.0–7.0 cm
Flower	Rosaceous	0.9–1.0 cm

Spot Identification
*Lamina: subulate with apical notch.
**Inflorescence: axillary raceme.
***Corolla: petals white

TABLE 5.25 Comparative Features Among Two Species of *Lumnitzera*

Habit/organ	Attribute	Two species of Lumnitzera	
		L. littorea	**L. racemosa**
Aerial roots	Form	○ Absent	○ Looping
Inflorescence	Form	○ Terminal raceme	○ Axillary raceme
Corolla	Color	● Dark red	● White

● denoting salient feature, ○ denoting common feature.

5.3.20 *NYPA FRUTICANS* WURMB (*ARECACEAE*)

(Verh. Batav. Genootsch. Kunsten 1: 349–1779.)
RLC&C: LC (ver 3.1); YP: 2010; DA: 2008–03–07
E. – Mangrove palm; VN: *B. – Gol-pata*

Diagnostic description (Figures 5.39 and 5.40; Table 5.26)

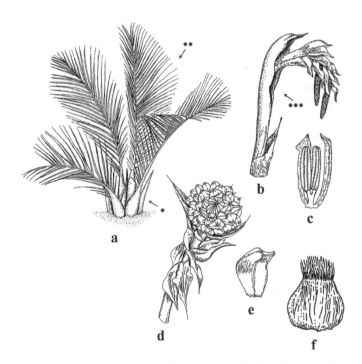

FIGURE 5.39 (a) Canopy; (b) male inflorescence (×1/10); (c) androecium (×1/2); (d) female inflorescence (×1/8); (e) carpel (×1/2.5); (f) fruit (×1/10).

FIGURE 5.40 (See color insert.) (a) Canopy; (b) male inflorescence; (c) female inflorescence; (d) fruits with incipient vivipary.

A palm looks like sunken coconut plant with rhizomatous subterranean stem; fibrous roots; leaf base sheathing enclosing stem, usually submerged with oblique rising, pinnately compound; inflorescence spadix with long peduncle; male flowers covered by spathe; perianth: tepals 6, polytepalous; androecium: stamens 3, united to form central column; female flower enclosed by spathe; perianth: tepals 6, polytepalous; gynoecium: carpels 3, free; germination incipient vivipery.

TABLE 5.26 Salient Morphometry of *Nypa fruticans*

Habit/organ	Features	Size (range)
Lifeform and canopy	Rhizomatous palm, looks young coconut palm	4.0–5.0 m
Leaf (Form)	Sheathing base	4.0–5.0 m
Inflorescence	Spadix	40.0–60.0 cm
Fruit (Type)	Drupe	10.0–12.0× 20.0–25.0 cm

Spot Identification

*Palm: looks like sunken coconut plant with rhizomatous stem.
**Leaf: pinnately compound.
***Inflorescence: spadix.

5.3.21 *PHOENIX PALUDOSA* ROXB. *(ARECACEAE)*

(Fl. Ind. ed. 1832–3: 789–1832.)
RLC&C: NT (ver 3.1); YP: 2010; DA: 2008–03–07
E. – Mangrove Date palm; VN: B. – Hental

Diagnostic description (Figures 5.41 and 5.42; Tables 5.27)

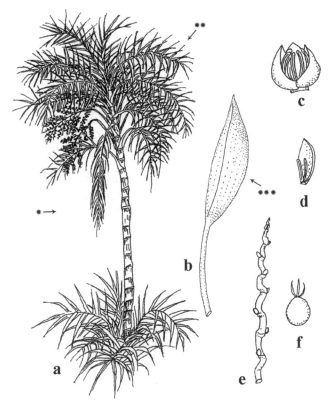

FIGURE 5.41 (a) Canopy; (b) spathe (×1/10); (c) male flower (×1/1.5); (d) tepal with stamens (×1/2); (e) part of female inflorescence (×1/5); (f) carpel (×1/3).

FIGURE 5.42 **(See color insert.)** (a) Canopy; (b) twig with fruits.

An unbranched palm forming bushy vegetation; fibrous roots; sometimes aerial pneumatothods also develop; leaves pinnately compound, spirally arranged on the top of stem, basal foliages modified into sharp spines, upper foliages with sheathing base; inflorescence spadix with long yellow peduncle; male flowers enclosed by spathe, sessile; calyx: sepals 3, gamosepalous; corolla: petals 3, polypetalous, yellow; androecium: stamens 6, free, sessile; female flower enclosed by spathe, sessile; calyx: sepals 3, gamosepalous; corolla: petals 3, polypetalous, yellow; gynoecium: carpels 3, syncarpous, stigma 3.

TABLE 5.27　Salient Morphometry of *Phoenix paludosa*

Habit/organ	Features	Size (range)
Lifeform and canopy	Unbranched palm	4.0–5.0 m
Leaf (Form)	Rachis	1.5–2.0 m
Inflorescence	Spadix	30.0–45.0 cm

Spot Identification
*Palm: An unbranched palm forming bushy vegetation.
**Leaf: pinnately compound, 1.5–2.0 m long.
***Inflorescence: male flowers enclosed by spathe.

5.3.22 *RHIZOPHORA APICULATA* BLUME (RHIZOPHORACEAE)

(Enum. Pl. Javae 1: 91–1827.)
RLC&C: LC (ver 3.1); YP: 2010; DA: 2008–03–07
VN: B. – Garjan

Diagnostic description (Figures 5.43 and 5.44; Tables 5.28 and 5.31)

Small tree; aerial roots forming slanting architecture, known as stilt roots, arising from trunk; leaves simple, petiolate, decussate; lamina ovate-lanceolate, acute; inflorescence cyme with a pair of flowers (2) at each peduncle; flowersessile; calyx: sepals 4, polysepalous; corolla: petals 4, polypetalous; androecium: stamens 11–12, free; gynoecium: carpels 2, syncarpous; germination vivipary, hypocotyl, each with rod-shaped and blunt end, hanging down from branches.

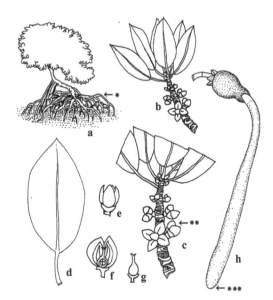

FIGURE 5.43 (a) Canopy; (b) flowering twig (×1/5); (c) twig with inflorescence (×1/5); (d) leaf (×1/3); (e) flower (×1/2); (f) section of flower (×1/1.5); (g) carpel (×1/1.5); (h) hypocotyl (×1/6).

FIGURE 5.44   **(See color insert.)** (a) Canopy; (b) twig with flower buds; (c) twig with hypocotyls.

TABLE 5.28   Salient Morphometry of *Rhizophora apiculata*

Habit/organ	Features	Size (range)
Lifeform and canopy	Tree	5.0–12.0 m
Root (system)	Stilt (Aerial)	0.5–2.5 m
Lamina (shape)	Ovate-lanceolate	10.0–13.0 cm ×5.0–6.0 cm
Inflorescence (type)	Cyme	2.2–2.5 cm
Flower	Sessile	0.8–1.2 cm
Hypocotyls (length)	Rod-shaped	45–55 cm

Key Characters for Spot Identification
*Stilt root: slanting architecture.

**Flower: 2 numbers at pair, below leafy cluster.

***Hypocotyl: 45–55 cm long, smooth surface, almost blunt end.

5.3.23 *RHIZOPHORA MUCRONATA* LAM. *(RHIZOPHORACEAE)*

(Encycl. 6: 189–1804.)

RLC&C: LC (ver 3.1); YP: 2010; DA: 2008–03–07

E. – Mangrove; VN: B. – Garjan, Bhora; Bo. – Kandel; Kan. – Kandale; Mal. – Pikantal; Tam. – Kandal; Tel. – Uppuponna

Diagnostic description (Figures 5.45 and 5.46; Tables 5.29 and 5.31)

Small tree; aerial roots forming bow architecture, known as stilt roots, arising from trunk; leaves simple, petiolate, decussate; lamina ovate-elliptic, mucronate; inflorescence cyme with 2 pair of flowers (4) at each peduncle; flowers pedicellate; calyx: sepals 4, polysepalous; corolla: petals 4, polypetalous; androecium: stamens 8, free; gynoecium: carpels 2, syncarpous; germination vivipary, hypocotyl, each with rod-shaped with cork wart on surface and tapering end, hanging down from branches.

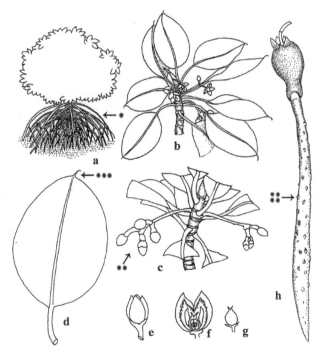

FIGURE 5.45   (a) Canopy; (b) flowering twig (×1/7); (c) twig with inflorescence (×1/5); (d) leaf (×1/5); (e) flower (×1/2.5); (f) section of flower (×1/2.5); (g) carpel (×1/3); (h) hypocotyl (×1/10).

FIGURE 5.46   (See color insert.) (a) Canopy; (b) twig with flower buds; (c) twig with hypocotyls.

TABLE 5.29   Salient Morphometry of *R. mucronata*

Habit/organ	Features	Size (range)
Lifeform and canopy	Tree	10.0–15.0 m
Root (system)	Stilt (Aerial)	0.5–3.0 m
Lamina (shape)	Ovate-elliptic	15.0–18.0 cm× 7.0–9.0 cm
Inflorescence (type)	Cyme	4.0–5.5 cm
Peduncle (length)	Long	2.5–3.0 cm
Hypocotyl (length)	Rod-shaped	60.0–75.0 cm

Key Characters for Spot Identification
*Stilt root: bow architecture.
**Flower: 4 numbers in each inflorescence.
***Leaf apex: mucronate.
****Hypocotyl: 60–75 cm long, warty surface.

5.3.24 RHIZOPHORA STYLOSA GRIFF. (RHIZOPHORACEAE)

(Not. Pl. Asiat. 4: 665–1854.)
RLC&C: LC (ver 3.1); YP: 2010; DA: 2008–03–07

Diagnostic description (Figures 5.47 and 5.48; Tables 5.30 and 5.31)

Small tree; aerial roots forming slanting architecture, known as stilt roots, arising from trunk; stem scars conspicuous; leaves simple, petiolate, decussate; lamina ovate-lanceolate; inflorescence cyme, 4 flowers at each peduncle; flowers, pedicellate, short; calyx: sepals 4, polysepalous; corolla: petals 4, polypetalous; androecium: stamens 8, free; gynoecium: carpels 2, syncarpous, style long; germination vivipary, hypocotyl, each with rod-shaped with few cork warts on surface and tapering end, hanging down from branches.

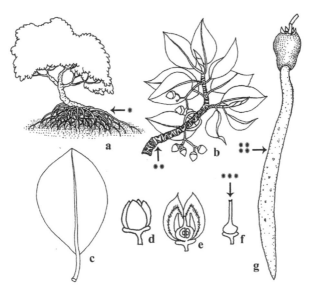

FIGURE 5.47 (a) Canopy (b) flowering twig (×1/4) (c) leaf (×1/3) (d) flower (×1/2); (e) section of flower (×1/1.5); (f) carpel (×1/1.5); (g) hypocotyl (×1/4).

FIGURE 5.48   **(See color insert.)** (a) Canopy; (b) flower showing distinct style.

TABLE 5.30   Salient Morphometry of *R. stylosa*

Habit/organ	Features	Size (range)
Lifeform and canopy	Tree	8.0–12.0 m
Root (system)	Stilt (Aerial)	0.5–2.5 m
Lamina (shape)	Ovate-lanceolate	8.0–10.0 cm × 4.5–5.0 cm
Inflorescence	Cyme	3.5–4.5 cm
Style (shape)	Straight	0.4–0.5 cm
Hypocotyl (Length)	Rod-shaped	20.0–40.0 cm

Key Characters for Spot Identification
* Stilt root: slantingarchitecture.
** Stem: conspicuous scars, inflorescence with 4 flowers.
*** Style: 0.4–0.5 cmlong.
****Hypocotyl:20–40 cm long, warty obscure.

TABLE 5.31   Comparative Identification among Three Species of *Rhizophora*

Habit/organ	Attribute	Three species of Rhizophora		
		R. apiculata	R. mucronata	R. stylosa
Stem	Scars (surface)	○ seen	○ seen	● conspicuous
Leaf	Apex	○ acute	● mucronate	○ acute/mucronate
Flower	Number	● 2 (Pair)	○ 4	○ 4
	Stalk	● sessile	○ pedicelate	○ pedicelate
	Petal	● smooth	○ pubescent	○ pubescent
	Style	○ sessile	○ sessile	● long
Hypocotyl	Length (cm)	● 45–55	● 60–75	● 20–40
	Surface (warty)	○ smooth	● distinct	○ occasional
	Apex	● blunt	○ tapering	○ tapering

● denoting salient feature, ○denoting common feature.

5.3.25   *SCYPHIPHORA HYDROPHYLACEA* C.F. GAERTN. (RUBIACEAE)

(Suppl. Carp. 91–1806.)
RLC&C: LC (ver 3.1); YP: 2010; DA: 2008–03–07
VN: B. – Tagri ban
[Note: The monotypic mangrove genus *Scyphiphora* was first described by Gaertner in 1805 as *Scyphiphora hydrophylacea*. But Tomlinson (1986) uses *Scyphiphora hydrophyllacea* in his book, 'The botany of mangroves,' on pages 361–364. Gaertner (1805) used the spelling hydrophylacea in his original description, but hydrophyllaceae is orthographically correct (from hydro, water, and phyllon, leaf). However, we prefer to retain the specific epithet *hydrophylacea* since Gaertner name is used in author citation as *Scyphiphora hydrophylacea* C. F. Gaertner]

Diagnostic description (Figures 5.49 and 5.50; Table 5.32)

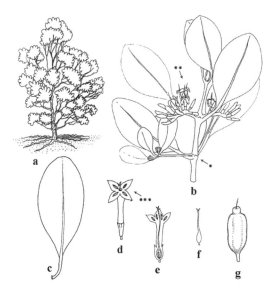

FIGURE 5.49   (a) Canopy; (b) flowering twig (×1/4); (c) leaf (×1/2); (d) flower (×1/1.5); (e) section of flower (×1/1.5); (f) carpel (×1/1.5); (g) fruit (×1/2).

FIGURE 5.50   (See color insert.) (a) Canopy; (b) inflorescence; (c) fruits.

Shrub; profusely branching; aerial roots slightly wavy, occasionally found; nodes ridged; leaves simple, decussate, interstipulate; lamina ovate-elliptic, obtuse; inflorescence condensed cyme; flowers tetramerous; calyx: sepals 4, gamosepalous, forming a tube; corolla: petals 4, white, gamopetalous; androecium: stamens 4, inserted, epipetalous; gynoecium: ovary one locule, stigma bifurcated.

TABLE 5.32   Salient Morphometry of *Scyphiphora hydrophylacea*

Habit/organ	Features	Size (range)
Lifeform and canopy	Shrub, under canopy	3.0–4.0 m
Lamina (shape)	Ovate-elliptic	6.0–10.0 cm × 3.0–5.5 cm
Inflorescence	Condensed raceme	2.0–2.5 cm
Flower	Rosaceous	1.0–1.2 cm × 0.3–0.5 cm

Spot Identification
*Stem: ridged node with interstipulate.
**Inflorescence: condensed raceme.
***Flower: tetramerous with white petal.

5.3.26 *SONNERATIA ALBA* SM. (LYTHRACEAE/ SONNERATIACEAE)

(Cycl. 33(I): Sonneratia no. 2–1816.)
RLC&C: LC (ver 3.1); YP: 2010; DA: 2008–03–07
VN: B. – Chak kaora

Diagnostic description (Figures 5.51 and 5.52; Tables 5.33 and 5.38)

Shrub to small tree; presence of slender aerial breathing roots known as pneumatophore; leaves simple, petiolate, decussate; lamina ovate-elliptic, obtuse; inflorescence cyme, solitary or 2–3 flowers at one peduncle; flower pedicellate; calyx: sepals 6; corolla: petals 6, polypetalous, white, inconspicuous; androecium: stamens indefinite, free; gynoecium: ovary syncarpous; fruit green, round flat with 6 sepal teeth, numerous seeds.

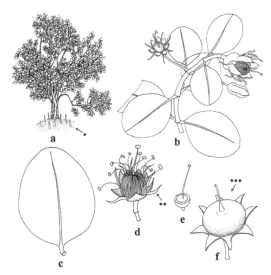

FIGURE 5.51   (a) Canopy; (b) flowering twig (×1/5); (c) leaf (×1/2); (d) flower (×1/3); (e) carpel (×1/3); (f) fruit (×1/5).

FIGURE 5.52   (**See color insert.**) (a) Canopy; (b) flowering twig.

TABLE 5.33 Salient Morphometry of *Sonneratia alba*

Habit/organ	Features	Size (range)
Lifeform and canopy	Bushy shrub	5.0–7.0 m
Aerial roots	Pneumatophores	20–50 × 1.5–2.0 cm
Lamina (shape)	Ovate-elliptic	8.0–12.0 cm × 4.0–5.0 cm
Inflorescence	Cyme	4.0–6.0 cm
Flower		4.0–6.0 cm × 4.0–6.0 cm
Fruit (Type)	Berry	4.5–7.5 cm (diameter)

Spot Identification

*Aerial roots: pneumatophores.

**Corolla: white but inconspicuous.

***Fruit: fruit round flat with 6 sepal teeth.

5.3.27 *SONNERATIA APETALA* BUCH.-HAM. *(LYTHRACEAE/ SONNERATIACEAE)*

(Embassy Ava 477–1800.)

RLC&C: LC (ver 3.1); YP: 2010; DA: 2008–03–07

VN:B. – Tak-Keora

Diagnostic description (Figures 5.53 and 5.54; Tables 5.34 and 5.38)

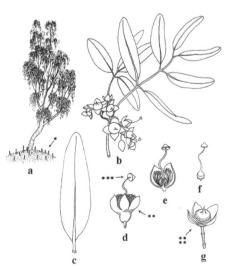

FIGURE 5.53 (a) Canopy; (b) flowering twig (×1/5); (c) leaf (×1/2.5); (d) flower (×1/1.5); (e) section of flower (×1/1.5); (f) carpel (×1/1.5); (g) fruit (×1/3).

FIGURE 5.54   **(See color insert.)** (a) Canopy; (b) flower; (c) a bunch of fruits.

Tall tree with drooping branches; bark light brown; presence of slender aerial breathing roots known as pneumatophore; leaves simple, petiolate, decussate; lamina oblong, obtuse; inflorescence terminal cyme, 3–7 flowers at one peduncle; flower rosaceous, pedicellate; calyx: sepals 4; corolla: petals absent; androecium: stamens many, free; gynoecium: carpels 4, syncarpous, stigma looks flat umbrella; fruit green, round with 4 sepal teeth.

TABLE 5.34   Salient Morphometry of *S. apetala*

Habit/organ	Features	Size (range)
Lifeform and canopy	Tall tree	15.0–30.0 m
Aerial roots	Pneumatophores	75.0–150 cm × 8.0 cm × 1.5–2.0 cm
Lamina (shape)	Oblong	8.0–12.0 cm × 1.5–3.0 cm
Inflorescence	Cyme	2.5–4.0 cm
Flower	Rosaceous	1.2–2.5 cm × 1.5–2.0 cm
Fruit (Type)	Berry	1.5–2.0 cm (diameter)

Spot Identification
*Aerial roots: pneumatophores.
**Corolla: absent.
***Stigma: flat umbrella.
****Fruit: fruit round with 4 sepal teeth.

5.3.28 *SONNERATIA CASEOLARIS* (L.) ENGL. *(LYTHRACEAE/ SONNERATIACEAE)*

(Nat. Pflanzenfam. Nachtr. 1: 261–1897.)
RLC&C: LC (ver 3.1); YP: 2010; DA: 2008–03–07
VN: B. – Chak-Keora; Bo. – Chipi; Tam. – Kinnai; Kan. – Kandale; Mal. – Thirala; Ma. – Tiwar; Odi. – Sundarignua

Diagnostic description (Figures 5.55 and 5.56; Tables 5.35 and 5.38)

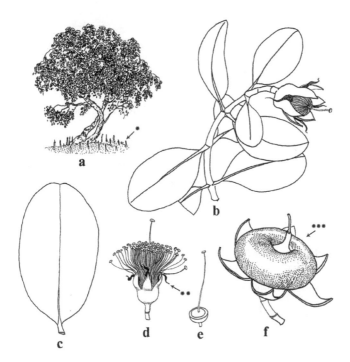

FIGURE 5.55 (a) Canopy; (b) flowering twig (×1/5); (c) leaf (×1/2); (d) flower (×1/3); (e) carpel (×1/3); (f) fruit (×1/5).

FIGURE 5.56   **(See color insert.)** (a) Canopy; (b) flowering bud; (c) fruit.

Small tree; bark pale brown, smooth, fissured at maturity; presence of slender aerial breathing roots known as pneumatophore; leaves simple, petiolate, decussate; lamina ovate-elliptic, obtuse; inflorescence cyme, solitary or 2–3 flowers at one peduncle; flowers pedicellate, showy; calyx: sepals 6; corolla: petals 6, polypetalous, dark pink; androecium: stamens indefinite, free, filaments white; gynoecium: ovary syncarpous, stigma capitate; fruit green, round flat with 6 sepal teeth.

TABLE 5.35   Salient Morphometry of *S. caseolaris*

Habit/organ	Features	Size (range)
Lifeform and canopy	Tree	6.0–8.0 m
Aerial roots	Pneumatophores	75.0–150 cm × 8.0 cm × 1.5–2.0 cm
Lamina (shape)	Ovate-elliptic	8.0–12.0 cm × 4.0–5.0 cm
Inflorescence	Cyme	4.0–6.0 cm
Flower	Rosaceous	4.0–6.0 cm × 4.0–6.0 cm
Fruit (Type)	Berry	5.0–8.0 cm (diameter)

Spot Identification
*Aerial roots: pneumatophores.
**Corolla: petals dark pink.
***Fruit: fruit round flat with 6 sepal teeth.

5.3.29 *SONNERATIA GRIFFITHII KURZ (LYTHRACEAE/ SONNERATIACEAE)

(Prelim. Rep. Forest Pegu App. B: 54–1875.); *treated as unresolved name as per the web
(www.theplantlist.org)
RLC&C: CR (ver 3.1); YP: 2010; DA: 2008–03–07
VN: B. – Ora

Diagnostic description (Figures 5.57 and 5.58; Tables 5.36 and 5.38)

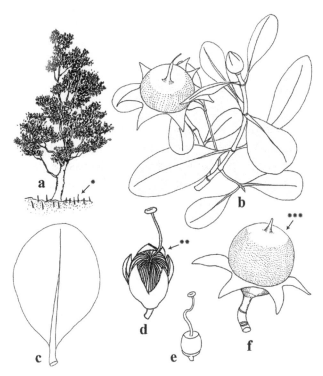

FIGURE 5.57 (a) Canopy; (b) flowering twig (×1/5); (c) leaf (×1/2); (d) flower (×1/3); (e) carpel (×1/3); (f) fruit (×1/5).

FIGURE 5.58 **(See color insert.)** (a) Canopy; (b) flowering bud; (c) fruits.

Medium to tall tree; bark deep brownish, young bark smooth but mature bark fissured; presence of slender aerial breathing roots known as pneumatophore; leaves simple, petiolate, decussate; lamina obovate, round; inflorescence solitary cyme; flower pedicellate; calyx: sepals 6–8; corolla: petals present at bud but abscise at maturity; androecium: stamens numerous, free; gynoecium: ovary syncarpous, stigma round globose; fruit green, round slightly flat with 6–8 sepal teeth.

TABLE 5.36   Salient Morphometry of *S. griffithii*

Habit/organ	Features	Size (range)
Lifeform and canopy	Tall tree	10.0–20.0 m
Aerial roots	Pneumatophores	50.0–80.0 cm × 1.0–1.5 cm
Lamina (shape)	Obovate	5.0–10.0 cm × 3.0–6.0 cm
Inflorescence	Cyme	4.0.0–5.0 cm
Flower	Rosaceous	3.0–5.0 cm × 5.0–7.0 cm
Fruit (Type)	Berry	5.0–7.0 cm (diameter)

Spot Identification
*Aerial roots: pneumatophores.
**Corolla: petals absent at maturity.
***Fruit: fruit round slightly flat with 6–8 sepal teeth.

5.3.30 *SONNERATIA OVATA* BACKER *(LYTHRACEAE/ SONNERATIACEAE)*

(Bull. Jard. Bot. Buitenzorg, sér. 3, 2: 329–1920.)
RLC&C: NT (ver 3.1); YP: 2010; DA: 2008–03–07

Diagnostic description (Figures 5.59 and 5.60; Tables 5.37 and 5.38)

Medium to tall tree; light gray bark, fissured; presence of stout pneumatophores with pointed end; leaves simple, petiolate, decussate; lamina surface slightly corrugated, broadly ovate, obtuse; inflorescence solitary or three flowers a group; flower globose; calyx: sepal 6; corolla: petal absent; androecium: stamens numerous, white; gynoecium: stigma capitate; fruit green, globose, grooved surface with 6 sepal teeth.

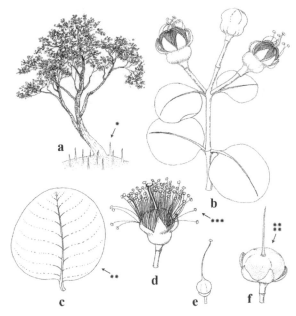

FIGURE 5.59 (a) Canopy; (b) flowering twig (×1/5); (c) leaf (×1/2); (d) flower (×1/3); (e) carpel (×1/3); (f) fruit (×1/5).

FIGURE 5.60   (See color insert.) (a) Canopy; (b) pneumatophores; (c) leaf; (d) flower; (e) fruit.

TABLE 5.37   Salient Morphometry of *S. ovata*

Habit/organ	Features	Size (range)
Lifeform and canopy	Medium tree	7.0–10.0 m
Aerial roots	Pneumatophores	120.0–200.0 cm
Lamina (shape)	Ovate	5.0–9.0 cm × 4.0–8.6 cm
Inflorescence	Cyme	4.5–5.0 cm
Flower	Rosaceous	3.0–5.0 cm × 5.0–7.0 cm
Fruit (Type)	Berry	4.0–6.0 cm ×3.5–4.5 cm (diameter)

Spot Identification
*Aerial roots: pneumatophores stout and sharply pointed.
**Leaf: lamina broadly ovate and surface corrugated.
***Androecium: stamen numerous and brightly white.
****Fruit: green, globose and grooved surface.

TABLE 5.38   Comparative Identification among Five Species of *Sonneratia*

Habit/ organ	Attribute	Five species of Sonneratia				
		S. alba	S. apetala	S. caseolaris	S. griffithii	S. ovata
Habit	Canopy	○Shrub	○Tree	○Tree	○ Tree	○ Tree
	Height (m)	○5–7	○15–30	○6–8	○ 10–20	○ 7–10
Leaf	Shape	○ ovate-elliptic	○ oblong	○ ovate-elliptic	○ obovate	○ broadly ovate
Calyx	Number	○6	●4	○6	○ 6–8	○ 6
Corolla	Petal	●White	○Absent	● Dark pink	○ Absent	○ Absent
Stamen	Color	● White	○ Creamy white	● Pinkish white	○ White pinkish	● Brightly white
Stigma	Shape	● flat umbrella	○ capitate	○ capitate	○ round globose	○ capitate
Fruit	Shape	○ round flat	○round	○round flat	○ round slight flat	● grooved surface

● denoting salient feature, ○denoting common feature.

5.3.31   *XYLOCARPUS GRANATUM* J. KOENIG *(MELIACEAE)*

(Naturforscher (Halle) 20: 2–1784.)
RLC&C: LC (ver 3.1); YP: 2010; DA: 2008–03–07
VN: Dhundul

Diagnostic description (Figures 5.61 and 5.62; Tables 5.39 and 5.41)

Medium tree; yellowish peeling in patches on bark; presence of well-developed aerial roots known as pneumatophores and developed root buttress; leaves petiolate, compound with 2, 4 or 6 leaflets, superimposed; leaflet ovate-elliptic; inflorescence mixed, irregularly branched; flowers pedicellate; epicalyx 4, free; calyx: sepals 4, polysepalous; corolla: petals 8, gamopetalous, creamy white; androecium: stamens 8, epipetalous; gynoecium: ovary syncarpous; fruit globose, woody fruit coat.

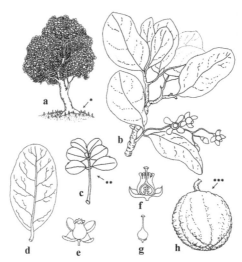

FIGURE 5.61   (a) Canopy; (b) flowering twig (×1/5); (c) compound leaf (×1/7); (d) leaflet (×1/2.5); (e) flower (×1/3); (f) section of flower (×1/3); (g) carpel (×1/2); (h) fruit (×1/8).

FIGURE 5.62   **(See color insert.)** (a) Canopy; (b) inflorescence; (c) twig with fruit.

TABLE 5.39 Salient Morphometry of *Xylocarpus granatum*

Habit/organ	Features	Size (range)
Lifeform and canopy	Medium tree	12.0–16.0 m
Aerial roots	Pneumatophores	40.0–60.0 cm
Leaflet (shape)	ovate-elliptic	8.0–11.0 cm × 4.5–5.5 cm
Inflorescence	Mixed	5.0–6.0 cm
Flower	Rosaceous	0.7–0.8 cm × 0.6–0.85 cm
Fruit	Round	> 15 cm (diameter)

Spot Identification

*Aerial roots: stout pneumatophores.
**Leaf: compound.
***Fruit: globose, woody fruit coat.

5.3.32 *XYLOCARPUS MOLUCCENSIS* (LAM.) M. ROEM. *(MELIACEAE)*

(Fam. Nat. Syn. Monogr. 1: 124–1846.)
RLC&C: LC (ver 3.1); YP: 2010; DA: 2008–03–07
VN: *Passur*

[Note:Tomlinson (1986) mentioned *Xylocarpus moluccensis* and *Xylocarpus mekongensis* as two different species. On the contrary, we identified *X. moluccensis* as an accepted name and *X. mekongensis* is treated as a synonym of former one as verified through online database as www.theplantlist.org – the working plant groups and related literatures.]

Diagnostic description (Figures 5.63 and 5.64; Tables 5.40 and 5.41)

Medium tree, dark brown bark, fissured; presence of well-developed pneumatophores; leaves petiolate, compound with 2, 4 or 6 leaflets, superimposed; leaflet ovate-oblong; inflorescence united cymes with mixed flowers; flower pedicellate; epicalyx 4, free; calyx: sepals 4 polysepalous; corolla: petals 8, gamopetalous, white; androecium: stamens 8, epipetalous; gynoecium: ovary syncarpous; fruit globose with green fruit coat.

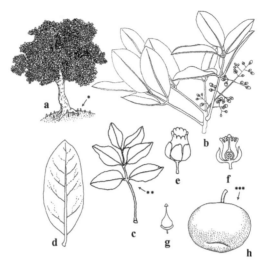

FIGURE 5.63   (a) Canopy; (b) flowering twig ($\times 1/5$); (c) compound leaf ($\times 1/7$); (d) leaflet ($\times 1/2.5$); (e) flower ($\times 1/3$); (f) section of flower ($\times 1/3$); (g) carpel ($\times 1/2$); (h) fruit ($\times 1/6$).

FIGURE 5.64   (See color insert.) (a) Canopy; (b) inflorescence; (c) fruits.

TABLE 5.40 Salient Morphometry of *X. moluccensis*

Habit/organ	Features	Size (range)
Lifeform and canopy	Medium tree	12.0–20.0 m
Aerial roots	Pneumatophores	40.0–60.0 cm
Leaflet (shape)	Ovate-elliptic	10.0–14.0 cm × 4.5–5.5 cm
Inflorescence	Mixed	5.0–8.0 cm
Flower	Rosaceous	0.7–0.8 cm × 0.3–0.5 cm
Fruit	Round	< 10 cm (diameter)

Spot Identification

*Aerial roots: stout pneumatophores.

**Leaf: compound.

***Fruit: globose, with green fruit coat.

TABLE 5.41 Comparative Features among Two Species of *Xylocarpus*

Habit/organ	Attribute	Two species of Xylocarpus	
		X. granatum	X. moluccensis
Leaflet	Shape	○ ovate-elliptic	○ ovate-oblong
Corolla	Color	● woody fruit coat	● soft fruit coat

● denoting salient feature, ○ denoting common feature.

5.4 MANGROVE ASSOCIATES

Mangrove associates include 14 species arranged alphabetically, with each species described concisely.

5.4.1 *ACROSTICHUM AUREUM* L. *(PTERIDACEAE)*

(Sp. Pl. 2: 1069–1753.)
RLC&C: LC (ver 3.1); YP: 2010; DA: 2008–03–07
E. – Golden Leather fern; VN: B. – Hodo

Description (Figures 5.65 and 5.66)

Erect fern appears to be bushy shrub about 1.5 m high; rhizome globose with circinate vernation, young rachis covered with dense hair known as ramenta; fronds simple, semi-pinnatifid up to 1.0 m long × 4.0 cm width, mid-vein conspicuous; mature leaves with blunt tip, sporophyllous, diffused sporangia develop along the mid-vein at abaxial surface.

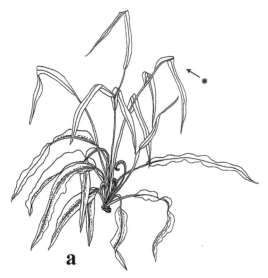

a

FIGURE 5.65   Canopy.

FIGURE 5.66   Canopy with sporophyllous leaves.

Spot Identification
*Leaf: semi-pinnatifid.

5.4.2   ACROSTICHUM SPECIOSUM WILLD. (PTERIDACEAE)

(Sp. Pl. 5(1–2): 117–1810.)
RLC&C: LC (ver 3.1); YP: 2010; DA: 2008–03–07

Description (Figure 5.67)

Erect fern appears to be bushy shrub about 1.0 m high, smaller than *Acrostichum aureum*; rhizome erect, stout, covered with large, broad, entire scales; fronds ascending, reddish/brown, pinnate, pinnae leathery, entire, 6–10 cm long × 2–4 cm width, dull green; mature leaves with pointed tip, sporophyllous, diffused sporangia develop along the mid-vein at abaxial surface.

FIGURE 5.67   Canopy.

Spot Identification
*Leaf: compound pinnate.

5.4.3 *AGLAIA CUCULLATA* (ROXB.) PELLEGR. *(MELIACEAE)*

(Fl. Indo-Chine 1: 771–1911.)

RLC&C: DD (ver 3.1); YP: 2010; DA: 2008–03–07

E. – Specific Maple; VN: B. – Amoora

Description (Figures 5.68 and 5.69)

Medium tree up to 15m high, occasionally grows up to 30m; aerial roots known as pneumatophores arising 60 cm high, along with plank buttress about 3.0m high; leaves compound, 2–4pairs; leaflet, oblong-elliptic, 7.5–10.0 cm× 3.0–6.0 cm, acute; inflorescence catkin, pendulous; flower small; calyx: sepals rounded; corolla: petals oblong; androecium: staminal tube shorter than petals; gynoecium: ovary stalked.

FIGURE 5.68 A twig with fruits.

FIGURE 5.69 (a) A twig with inflorescence; (b) close view of inflorescence.

Spot Identification
*Leaf: compound.
**Leaflet: oblong-elliptic.

5.4.4 ARDISIA ELLIPTICA THUNB. (PRIMULACEAE/ MYRSINACEAE)

(Nov. Gen. Pl. 119–1789.)

Description (Figure 5.70)

Bushy shrub up to 5.0 m high; leaves simple, spiral, petiole up to 1.0 cm long; lamina elliptic, 8.0–10.0 cm × 3.5–4.5 cm, inflorescence terminally condensed raceme, flowers cluster like umbel at apex, 1.0 cm diameter; flower pentamerous; calyx: sepals 5, rounded; corolla: petals 5, white, pointed; androecium: stamens 5; gynoecium: ovary globose with a simple style; fruit globose, 0.5 cm diameter, ripening black.

FIGURE 5.70   A twig with flowers & fruits.

Spot Identification
*Inflorescence: terminally condensed raceme.
**Fruit: globose and ripening black.

5.4.5   *BARRINGTONIA RACEMOSA* (L.) SPRENG. *(LECYTHIDACEAE)*

(Syst. Veg. 3: 127–1826.)

Description (Figure 5.71)

Medium tree up to 15 m high; bark blackish gray, fissured; leaves simple, petiolate, cluster at terminal shoot; lamina elliptic-oblong, 10.0–25.0 cm × 6.0–10.0 cm, acute; inflorescence terminal raceme, catkin, pendulous, looking attractive; flowers cream colored; stamens numerous, filament exerted; fruit rounded, ridged.

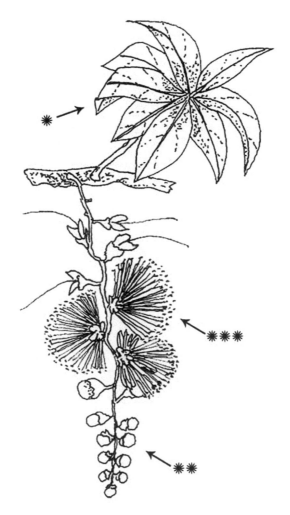

FIGURE 5.71 A twig with inflorescence.

Spot Identification
*Leaf: cluster at terminal shoot.
**Inflorescence: terminal raceme, catkin and pendulous
***Stamens: numerous, filament exerted.

5.4.6 *BROWNLOWIA TERSA* (L.) KOSTERM. *(MALVACEAE/ TILIACEAE)*

(Penerbitan Madj. Penget. Indonesia 1: 73–1959.)
RLC&C: NT (ver 3.1); YP: 2010; DA: 2008–03–07

Description (Figure 5.72)

Bushy shrub up to 3.0 m high, twining occasionally, twig appears to be brown; bark brownish black; leaves simple, alternate, petiole up to 2.0 cm long; lamina lanceolate, 10.0–14.0 cm × 2.5–4.0 cm, acute, pale brown at ventral surface; inflorescence mixed cyme, pendulous; flower pedicellate, 0.45 cm × 0.4 cm; calyx: sepals 4, polysepalous; corolla: petals 6, polypetalous; androecium: stamens numerous, free, filament exerted; gynoecium: carpels 3, apocarpous; fruit capsule, small.

FIGURE 5.72   A twig with inflorescence.

Spot Identification
*Leaf: pale brown at ventral surface.
**Inflorescence: mixed cyme and pendulous.

5.4.7   *CARALLIA BRACHIATA* (LOUR.) MERR. *(RHIZOPHORACEAE)*

(Philipp. J. Sci. 15: 249–1919.)

Description (Figure 5.73)

Tall tree up to 25 m high, occasional growing to 35 m; bark brown; aerial roots forming stilt roots, along with root buttress; leaves simple, decussate, petiolate; lamina ovate-lanceolate, acuminate, 9.0–15.0 cm × 4.5–7.5 cm; inflorescence axillary; flower pedicellate, 0.4–0.6 cm ×0.8–1.0 cm; calyx: sepals free, reddish at base; corolla: petals 5–8; androecium: stamens 10–16, free; gynoecium: ovary syncarpous; fruit berry, small, round, 0.5–1.0 cm × 0.5–1.0 cm; seeds 10–20 numbers.

FIGURE 5.73   A twig with inflorescence.

Spot Identification
*Inflorescence: axillary.

5.4.8   *CERBERA ODOLLAM* GAERTN. *(APOCYNACEAE)*

(Fruct. Sem. Pl. 2: 193–1791.)

VN:B. – Babur; Tam. – Kattarali; Ma. – Utalam; Ma. – Sukanu

Description (Figures 5.74 and 5.75)

Shrub to small tree, 8.0–12.0 m high; distinct leaf scars on branch lets; leaves simple, petiolate, decussate; lamina lanceolate, acuminate, 10.0–15.0 cm × 4.0–5.5 cm; inflorescence axillary with many flowers in cluster, 8.0–15.0 cm long; flower bell-shaped; calyx: sepals 5, free; corolla: petals white, forming long tube at base, yellow eye at center; androecium: stamens epipetalous, inserted, 0.9–1.2 cm long; gynoecium: ovary syncarpous; fruit spherical, 4.5–6.0 cm × 4.0–5.5 cm, green.

FIGURE 5.74   A twig with inflorescence.

FIGURE 5.75 (a) A flowering twig; (b) fruit.

Spot Identification
*Corolla: petals white, yellow eye at center

5.4.9 *CLERODENDRUM INERME (L.) GAERTN. (VERBENACEAE)*

(Fruct. Sem. Pl. 1: 271–1788.), *considered as a synonym of *Volkameria inermis* L. (Lamiaceae),
Sp. Pl. 637–1753, as per the web (www.theplantlist.org)
VN:B. – *Bonjui;* H.-*Lanjai;* S. – *Kundali;* Bo. – *Vanajai;* Mal. – *Nirnochi;* Tam. – *Pinarichanganguppi; T*

Description (Figures 5.76 and 5.77)

Straggling bushy shrub up to 3.0 m high; leaves simple, decussate, petiole 0.6–1.0 cm long; lamina ovate-elliptic, 3.5–4.5 cm × 1.5–2.0 cm, entire, acute; inflorescence axillary cyme, 3.5–4.5 cm long; flower bell-shaped, 5.0–5.6 cm ×0.8–1.0 cm; calyx: sepals 5, polysepalous; corolla: petal 5, 1.5–2.0 cm × 0.1–0.2 cm, forming long tube, white; androecium: stamens 5, epipetalous, filaments long extruded; gynoecium: carpels 2, style up to 4.5 cm long; fruit pyriform, drupe.

FIGURE 5.76   A twig with inflorescence.

FIGURE 5.77 A flowering twig.

Spot Identification
*Inflorescence: axillary cyme.
**Filaments: long extruded.

5.4.10 CYNOMETRA IRIPA KOSTEL. *(LEGUMINOSAE)*

RLC&C: LC (ver 3.1); YP: 2010; DA: 2008–03–07
VN:B. – Shingra; S. – Madhuka; Mal. – Irripa; Tam. – Irudbu

Description (Figure 5.78)

Small tree up to 8.0 high; bark gray, glabrous; leaves unijugate or bijugate, compound, paripinnate with 2 pairs of leaflets, upper pair larger than lower pair; leaflet obliquely oblong or obovate-oblong, 4.0–8.0 cm × 1.5–4.0 cm, obtuse, base asymmetrically acute; inflorescence raceme; flowers white to reddish-purple; calyx: sepals 3–5; corolla: petals 5, free; androecium: stamens 10, free; gynoecium: ovary inserted, stigma capitates; fruit ellipsoid, 2.0–3.0 × 1.5–2.0 cm, entire surface wrinkled with lateral beak extending up to, 0.4–0.6 cm; single-seeded.

FIGURE 5.78   A twig with fruits.

Spot Identification
*Leaf: unijugate or bijugate, compound, paripinnate.
**Fruit: wrinkled surface with lateral beak.

5.4.11 *DOLICHANDRONE SPATHACEA* (L.F.) SEEM. *(BIGNONIACEAE)*

(J. Bot. 1: 226–1863.)

RLC&C: LC (ver 3.1); YP: 2010; DA: 2008–03–07

VN:B. – Gorshingiah; Tam. – Viribadiri; Mal. – Nirpponalyam

Description (Figures 5.79. and 5.80)

Medium tree up to 20 m high; bark gray or dark brown, fissured; leaf scars on terminal twigs; leaves decussate, compound, imparipinnate, 20.0–25.0 cm long, petiole up to 6.0 cm long; leaflets 2–4 pairs, ovate-lanceolate, 5.0–10.0 cm × 3.0–5.0 cm, mature leaflets turn reddish brown; inflorescence with 3–6 flowers in cluster at the terminal twig; flower large, zygomorphic; calyx: sepals green; corolla: petals tubular, 15.0–20.0 cm long, showy; androecium: stamens 4; gynoecium: ovary bilocular, 0.8–1.2 cm long; fruit capsule, pendulous up to 48.0 cm long, two valves.

FIGURE 5.79 A twig with flower.

FIGURE 5.80   (a) Flowers; (b) fruits.

Spot Identification
*Leaf: compound and imparipinnate.
**Flower: large with tubular petals up to 20.0 cm long.

5.4.12   *HIBISCUS TILIACEUS* L. *(MALVACEAE)*

Description (Figures 5.81 and 5.82)

Tall tree up to 20 m high, with mucilaginous latex; bark gray, fissured at maturity; leaves simple, alternate, inter-stipulate, stipules ovate; lamina cordate, 14.0–17.5 cm × 12.0–16.0 cm; inflorescence cyme, two flowers on each peduncle; flowers pentamerous, pedicel 2.0 cm long; epicalyx 5, free; calyx: sepals 5, bell-shaped; corolla: petals whitish yellow, crimson eye spot at center; androecium: stamen indefinite, monadelphous; gynoecium: carpels 5, ovary syncarpous, stigma 5, capitate; fruit capsule.

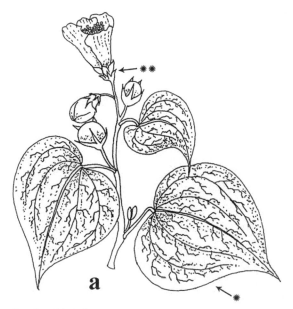

FIGURE 5.81 A twig with inflorescence.

FIGURE 5.82 A flowering twig.

Spot Identification
*Leaf: lamina cordate.
**Flower: epicalyx 5.

5.4.13 *PEMPHIS ACIDULA* J.R. FORST. & G. FORST. *(LYTHRACEAE)*

(Char. Gen. Pl. 34–1775.)
RLC&C: LC (ver 3.1); YP: 2010; DA: 2008–03–07

Description (Figure 5.83)

Small tree up to 9.5 m high; bark gray to brown, fissured at maturity; leaves simple, decussate, petiole short; lamina elliptic-lanceolate, fleshy, pubescent, 1.4–3.5 cm × 0.4–1.5 cm; flower axillary, solitary; calyx: sepals 5, base tubular, 0.5–0.8 cm, ribbed; corolla: petals 5, white, 0.4–0.6 cm × 0.3–0.4 cm; androecium: stamens 12; gynoecium: ovary syncarpous, stigma capitate; fruit capsule, spherical.

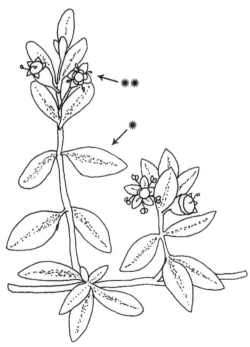

FIGURE 5.83   A twig with fruits.

Spot Identification
*Leaf: lamina fleshy, pubescent.
**Flower: petals white.

5.4.14 *THESPESIA POPULNEA* (L.) SOL. EX CORRÊA (MALVACEAE)

(Ann. Mus. Hist. Nat. 9: 290–1807.)
VN:B. – Paras; S. – Parisha; H &P.-Paraspipol; Bo. – Parsipu; Tam. – Puvarasu; Tel. – Gangaravi; Mal. –
Kallal; Kan. – Arasi

Description (Figures 5.84 and 5.85)

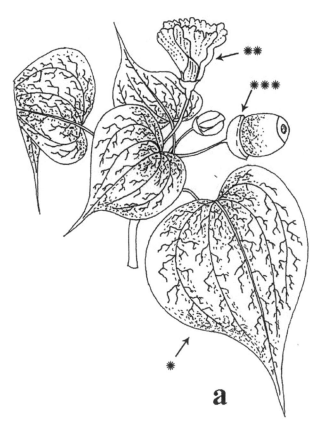

FIGURE 5.84 A twig with inflorescence and fruit.

FIGURE 5.85   A flowering twig.

Medium tree up to 15.0 m high, with mucilaginous latex; bark gray, fissured; leaves simple, alternate, intrastipulate, stipules lanceolate; petiole up to 10.5 cm long; lamina cordate, 12.0–15.0 cm × 9.0–11.7 cm, acuminate; inflorescence raceme, 2–4 flowers on each peduncle; flower pentamerous, 4.5–5.5 cm ×4.0–4.5 cm; calyx: sepals 5, cup-shaped structure; corolla: petals 5, showy, yellow but turn purple at maturity; red spot at center; androecium: stamens indefinite, monadelphous; gynoecium: ovary cup-shaped, stigma 5, capitate; fruit capsule with cup-shaped calyx, 2.5 cm × 2.5 cm.

Spot Identification
* Leaf: lamina cordate.
** Flower: petals yellow but turn purple at maturity, red spot at center.
*** Fruit: capsule with cup-shaped calyx.

5.5   MANGROVE HALOPHYTES

Mangrove halophytes include 18 species arranged alphabetically, with each species described concisely.

5.5.1 *ACANTHUS VOLUBILIS* WALL. *(ACANTHACEAE)*

(Pl. Asiat. Rar. 2: 56–1831.)
RLC&C: LC (ver 3.1); YP: 2010; DA: 2008–03–07
VN:B. – Lata harkoch

Description (Figures 5.86 and 5.87; Table 5.3)

Twining habit around the support, growing up to 4.0m high; leaves simple, decussate, petiole up to 1.0 cm long; lamina elliptic with wide middle, 6.0–8.0 cm × 4.5–5.6 cm, entire, obtuse; inflorescence raceme, peduncle up to 9.8 cm long; flower zygomorphic; calyx: sepals 4; corolla: petal 1, base form tube, terminally 4 short lobes, 2.0–2.5 cm × 1.2–1.6 cm, white; androecium: stamens 4, anthers free; gynoecium: carpels 2, syncarpous, stigma absent; fruit capsule up to 2.0 cm long.

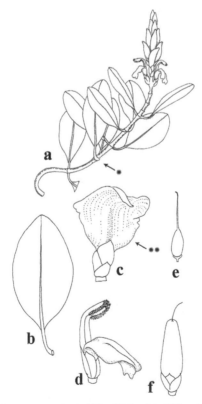

FIGURE 5.86 (a) A flowering twig (×1/5); (b) leaf (×1/3); *c*& (d) flower (×1/2.5); (e) carpel (×1/1.5); (f) fruit (×1/2).

FIGURE 5.87   (a) A twining plant; (b) inflorescence twig; (c) fruits.

Spot Identification
*Habit: twining.
**Flower: petals white.

5.5.2 *AELUROPUS LAGOPOIDES* (L.) THWAITES *(POACEAE)*

(Enum. Pl. Zeyl. 374–1864.)

Description (Figure 5.88)

Prostrate grass, creeping and proliferate on the ground; fibrous roots develop from nodes and anchor substrate; stem solid; leaf blade linear; inflorescence panicle, upright up to 0.8 cm high; flower pinkish, small.

FIGURE 5.88 A twig with inflorescence.

Spot Identification
*Habit: creeping.
**Flower: pinkish.

5.5.3 *CRINUM DEFIXUM KER GAWL. (AMARYLLIDACEAE)*

(J. Sci. Arts (London) 3: 105–1817.), *considered as a synonym of *Crinum viviparum* (Lam.) R. Ansari & V.J. Nair (*J. Econ. Taxon. Bot. 11: 205–1987 publ. 1988.*), as per the web (www.theplantlist.org)
RLC&C: LC (ver 3.1); YP: 2013; DA: 2010–05–03
E. – River Crinum Lily; VN:B. – Sukhadarsan; Bo. – Nagdown; M.-Vishamungil; Tel. – Kesarchettu

Description (Figure 5.89)

Herb with bulbous rhizome, growing up to 1.5m high; leaves simple, rosette, petiole sheathing at base; lamina linear, 80–100.0 cm × 10.0–12.0

cm, fleshy; inflorescence umbel, scape axillary, 5–10 flowers on each peduncle; flower large, showy, 4.5–5.5 cm × 4.5–5.0 cm; perianth: tepal 5, white; androecium: filament long, exerted up to 2.2 cm, anther extrude; gynoecium: ovary syncarpous, style long, navy green upper and white below.

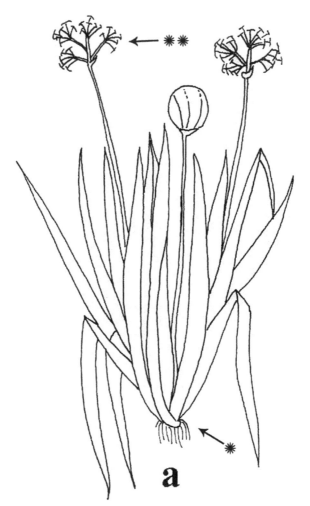

FIGURE 5.89   A plant with inflorescence.

Spot Identification
*Habit: herb with bulbous rhizome.
**Inflorescence: umbel, scape axillary, flower white.

5.5.4 *CRYPTOCORYNE CILIATA* (ROXB.) FISCH. EX WYDLER (*ARACEAE*)

(Linnaea 5: 428–1830.)
RLC&C: LC (ver 3.1); YP: 2013; DA: 2010–09–08

Description (Figures 5.90 and 5.91)

Herb erect up to 0.8 m high, forming dense bush; bulbous rhizome; leaves simple, rosette, petiole sheathing at base; lamina linear, 0.40–0.70 cm × 0.5–0.7 cm, fleshy; inflorescence with long stalk up to 0.4 cm high, remain hidden within canopy, flowers arranged terminally and remain closed.

FIGURE 5.90   A plant.

FIGURE 5.91   Partly submerged vegetation.

Spot Identification
*Habit: herb with bulbous rhizome.

5.5.5 *HALOSARCIA INDICA* (WILLD.) PAUL G. WILSON (AMARANTHACEAE)

(Nuytsia 3: 63–1980.): * *Salicornia brachiata* Miq. (Chenopodiaceae) is commonly known, but identified as a synonym of *H. indica* (Willd.) Paul G. Wilson *(Fl. Ned. Ind. 1: 1019–1858.)* as per the web (www.theplantlist.org) VN: Bo. – Machul; Mal & Tam. – Umari; Tel. – Koyyalu

Description (Figure 5.92)

Small herb up to 30.0 cm high, green, forming dense sprouting; stem succulent, jointed horizontal nodes and erect lateral branches; leaves simple, short petiole; lamina small, turn reddish at maturity; inflorescence greenish; flower small, not showy; fruit small, succulent with a single seed.

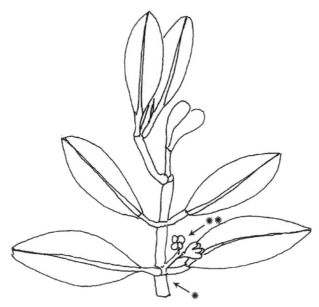

FIGURE 5.92   A twig with inflorescence.

Spot Identification
*Habit: succulent stem.
**Inflorescence: greenish, flower small.

5.5.6   *HELIOTROPIUM CURASSAVICUM* L. *(BORAGINACEAE)*

(Sp. Pl. 130–1753.)
VN: Nona hatisnur

Description (Figures 5.93 and 5.94)

Prostrate with branched herb, creeping on the ground; stem succulent, whitish green; leaves simple, decussate, sessile; lamina lanceolate, 1.5–2.0 cm × 0.3–0.5 cm, bluish, fleshy; inflorescence unipodial helicoid cyme, peduncle up to 15.0 cm high; flower small, 0.2–0.3 cm × 0.2–0.25 cm, sessile; calyx: sepals 5, polysepalous, succulent, persistent: corolla: petals 5, forming short hollow tube below, 5 pertite above, white; androecium: stamens 5, epipetalous; gamopetalous: carpels 2, ovary 4 chambered.

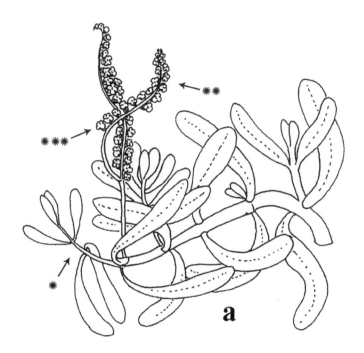

FIGURE 5.93   A twig with inflorescence.

FIGURE 5.94   A view of vegetation.

Spot Identification
*Habit: prostrate with branched herb.
**Inflorescence: unipodial helicoid cyme.
***Corolla: white.

5.5.7 *HOYA PARASITICA WALL. EX TRAILL (APOCYNACEAE/ ASCLEPIADACEAE)

(Trans. Hort. Soc. London 7: 23–1830.), *an unresolved name as per the web (www.theplantlist.org)

Description (Figure 5.95)

Twining herb; presence of milky latex; node ridged; leaves simple, decussate, petiolate; lamina ovate-elliptic, 4.0–6.0 cm × 2.0–3.0 cm, mucronate, base cup-shaped, fleshy; inflorescence cyme; flower pedunculate; calyx: sepals 5; corolla: petals 5; androecium: stamens 5, gynostegium; gynoecium: carpels 5, corona 5; fruits follicle, cylindrical, gradually tapering end.

FIGURE 5.95   A twig with fruits.

Spot Identification
*Habit: twining herb with milky latex.
**Fruit: follicle.

5.5.8 *HYDROPHYLAX MARITIMA* L.F. *(RUBIACEAE)*

(Suppl. Pl. 126–1782.)

Description (Figure 5.96)

Prostrate with profusely branching herb, creeping on the sandy beach; stem succulent; leaves simple, decussate, base sheathing partly, covering the stem; lamina ovate-elliptic, 2.0–2.5 cm ×1.5–1.7 cm, succulent, dark green; inflorescence axillary solitary cyme; flower large up to 3.4 cm long, tetramerous; calyx: sepals 4, gamosepalous; corolla, petals 4, gamopetalous, hollow tube below, 4 lobes above, purple; androecium: stamens 4, epipetalous, anthers violet; gynoecium: carpels 4, stigma absent; fruit achene, 2.2–2.5 cm ×0.8–1.0 cm.

FIGURE 5.96   A twig with flower and fruits.

Spot Identification
*Habit: prostrate with profusely branching herb.
**Flower: purple.
***Fruit: achene.

5.5.9 IPOMOEA PES-CAPRAE (L.) R. BR. (CONVOLVULACEAE)

(Narr. Exped. Zaire 477–1818.)

VN:B. – Chagal khuri; Bo. – Marjavel; H.-Dopatilata; Mal. – Atampa; S. – Sagaramekhala; Tam. – Attukkal; Tel. – Chivvulapillitige

Description (Figures 5.97 and 5.98)

Prostrate creeping herb, growing usually on sandy beach; internodes up to 11.0 cm long; leaves simple, alternate, petiole up to 12.0 cm long; lamina apically bilobed, each lobe broadly ovate, green, fleshy; inflorescence axillary solitary cyme, peduncle branched up to 7.0 cm long;flower pentamerous, 5.0–5.5 cm × 2.5–3.0 cm, calyx: sepals 5, polysepalous; corolla: petals 5, gamopetalous, funnel-shaped, violet at center; androecium: stamens 5, epipetalous, 2 large and 3 small, filament base covered by thick white velvet hairs; gynoecium: ovary syncarpous; fruit capsule with apical beak, 0.8–1.0 cm × 1.0–1.3 cm.

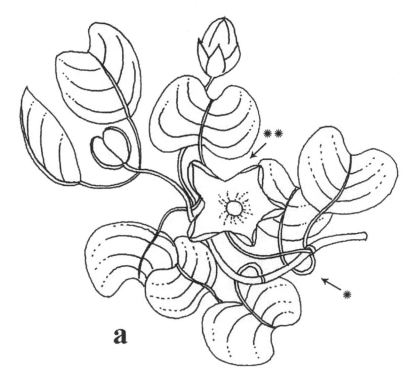

FIGURE 5.97 A twig with inflorescence.

FIGURE 5.98   A flowering twig.

Spot Identification
*Habit: prostrate herb with profuse branching
**Flower: corolla funnel-shaped, violet at center.

5.5.10 *MYRIOSTACHYA WIGHTIANA* (NEES EX STEUD.) HOOK.F. *(POACEAE)*

(Fl. Brit. India 7: 327–1896.)

Description (Figures 5.99 and 5.100)

Prostrate grass with erect branching; culms tufted, sometimes floats with long branching roots; leaves alternate, simple, petiole with basal long sheath, partly covering the culm; leaf blade linear-broad, serrated, acuminate; inflorescence panicle, rachis smooth, crowded, and whorled flowering, spikelet 4–8 flowers, compressed, pedicel short; fruit grain obliquely ovoid.

FIGURE 5.99 A twig with inflorescence.

FIGURE 5.100 Vegetation with inflorescence.

Spot Identification
*Habit: prostrate grass with erect branching.
**Inflorescence: panicle.

5.5.11 *PENTATROPIS CAPENSIS* (L. F.) BULLOCK *(APOCYNACEAE/ASCLEPIADACEAE)*

(Kew Bull. 10: 284–1955.)

Description (Figures 5.101 and 5.102)

Twining herb climbing around the support; node ridged; leaves simple, decussate, petiolate; lamina ovate-elliptic, 2.0–2.5 cm × 1.5–1.8 cm, mucronate, entire, base cup-shaped; inflorescence cyme; flower pentamerous, peduncule up to 2.5 cm long, numerous hairy bracts at base; calyx: sepals 5, corolla: petals 5, brownish-white; androecium: stamens 5, gynostegium, corona 5; gynoecium: ovary syncarpous; fruits follicle, cylindrical up to 4.0 cm long, gradually tapering end.

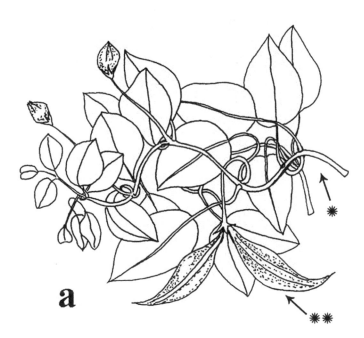

FIGURE 5.101   Twigs with inflorescence.

FIGURE 5.102   A twig with inflorescence.

Spot Identification
*Habit: twining herb.
**Fruit: follicle.

5.5.12 *PORTERESIA COARCTATA* (ROXB.) TATEOKA (POACEAE)

(Bull. Natl. Sci. Mus. Tokyo, B 8: 406–1965.)
VN:B. – Dhani Ghash

Description (Figures 5.103 and 5.104)

Grass with erect culm looking a vegetation of paddy, pseudo-tap root develops in loosely silted up soil, in consolidated soil strata a stout creeping rhizome spread horizontally with internode up to 14.0 cm long, branches grow upward from nodes; leaf sheath long, partly covering the culm; leaf blade linear, 20.0–30.0 cm × 0.4–0.5 cm, acuminate, margins dented; inflorescence spikelet up to 20.0 cm long; perianth: tepals 2, free, white; androecium: stamens 6, free; gynoecium: stigma feathery; fruit caryopsis, 1.2–1.4 cm × 0.4–0.45 cm.

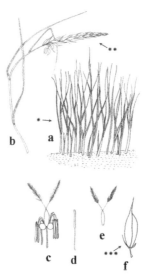

FIGURE 5.103   (a) A view of vegetation (×1/25); (b) flowering twig; (c) flower (×1/3); (d) stamen (×1/3); (e) carpel (×1/3); (f) fruit (×1/3).

FIGURE 5.104   (a) Landscape vegetation; (b) a close view of vegetative stand.

Spot Identification
*Habit: looking a vegetation of paddy.
**Inflorescence: spikelet.
***Fruits: caryopsis.

5.5.13 *SALVADORA PERSICA* L. *(SALVADORACEAE)*

(Sp. Pl. 122–1753.)
VN:B. – Brihatpilu; H&B.-Chotapilu; Bo. – Pilvu; P.-Pilu; Tam. – Perungoli; Tel. – Gogu

Description (Figure 5.105)

Shrub forming erect bush up to 5.0m high, with many drooping branches; leaves simple, opposite; lamina ovate-elliptic, 5.0–7.0 cm × 2.5–3.0 cm, slightly fleshy; inflorescence panicles up to 30.0 cm long; flower small, greenish; fruit berry, 0.8–1.0 cm long, fleshy, becoming red-scarlet after ripe.

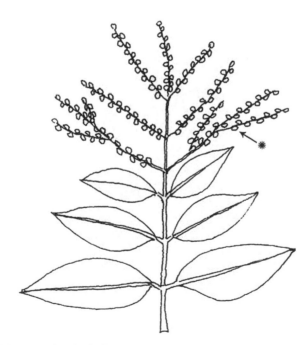

FIGURE 5.105   A twig with inflorescence.

Spot Identification
*Fruit: small berry, red-scarlet after ripe.

5.5.14 *SARCOLOBUS CARINATUS GRIFF. (APOCYNACEAE/ ASCLEPIADACEAE)

*(Not. Pl. Asiat. 4: 53–1854.),*an unresolved name as per the web (www. theplantlist.org)
VN: B. – Baole lata

Description (Figures 5.106 and 5.107)

FIGURE 5.106 A twig with inflorescence.

FIGURE 5.107 Plant with inflorescence.

Climbing herb, presence of milky latex; nodes slightly ridged but not jointed; leaves simple, decussate, petiole up to 2.7 cm long; lamina elliptic-oblong, 5.0–6.5 cm × 2.5–2.7 cm, acute; inflorescence corymbose up to 2.0 cm long, peduncle unbranched, bract scars distinct on the peduncle; flower small; calyx: sepals 5; corolla: petal 5, white-pink, scattered brown spots on outer surface of petal, 5 apical beak above and short hollow tube below; androecium: stamens 5, gynostegium; gynoecium: carpels 2, ovary pyramid shaped; fruit follicle with 2–3 flattened ribs and a short beak, 4.5–5.0 cm long.

Spot Identification
* Habit: climbing herb, presence of milky latex.
**Inflorescence: corymbose up to 2.0 cm long.

5.5.15 *SARCOLOBUS GLOBOSUS WALL. (APOCYNACEAE/ASCLEPIADACEAE)

(Asiat. Res. 12: 568–1816.), *an unresolved name as per the web (www.theplantlist.org)

VN: B. – Baole lata

Description (Figures 5.108 and 5.109)

Climbing herb, occasionally prostrate, presence of milky latex; nodes slightly ridged but not jointed; leaves simple, decussate, petiole up to 1.5 cm long; lamina elliptic-oblong, 7.5–8.5 cm × 3.5–4.0 cm, acute; inflorescence branched cyme, peduncle firstly bifurcate then trifurcate; flower small, 0.5 cm × 1.0 cm; calyx: sepals 5; corolla: petals 5, pink-white, dark pink at mid-portion, distinct hairy on both surfaces at mid-portion, white cilia attached to the inner surface at base; androecium: stamens 5, gynostegium; gynoecium: carpel 2, ovary pyramid shape; fruit follicle, brown globose with terminal beak, up to 5.0 cm long.

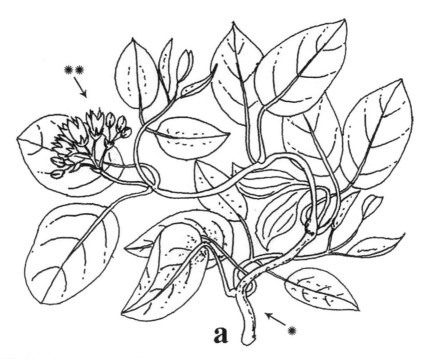

FIGURE 5.108　A twig with inflorescence.

FIGURE 5.109   A view of vegetation.

Spot Identification
*Habit: climbing herb, presence of milky latex.
**Inflorescence: branched cyme.

5.5.16   *SESUVIUM PORTULACASTRUM* (L.) L. *(AIZOACEAE)*

Syst. Nat. ed. 10–2: 1058–1759.
VN:B. – Gada bani

Description (Figures 5.110 and 5.111)

Prostrate herb with profuse branching; adventitious roots develop; stem fleshy, pink, nodes jointed; leaves simple up to 1.0 cm long, decussate, petiole up to 0.4 cm long; lamina lanceolate, 4.0–4.5 cm × 0.6–0.8 cm, acute; inflorescence solitary axillary cyme; flower showy, 0.7 cm × 0.7 cm; calyx: sepals 5, polysepalous, margin with pink; corolla absent; androecium: stamen numerous, free, pink-violet; gynoecium: carpels 3; fruit capsule with persistent calyx.

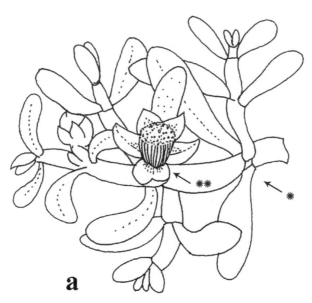

FIGURE 5.110   A twig with inflorescence.

FIGURE 5.111   Plant with flower.

Spot Identification
*Habit: prostrate herb with profuse branching.
**Flower: sepals 5, showy with pink margin.

5.5.17 SUAEDA MARITIMA (L.) DUMORT. (AMARANTHACEAE/CHENOPODIACEAE)

(Fl. Belg. 22–1827.)
VN: B. – Gire sak

Description (Figures 5.112 and 5.113)

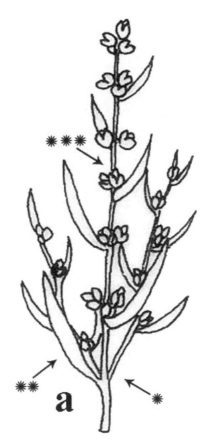

FIGURE 5.112 A twig with inflorescence.

FIGURE 5.113   (a) Partly submerged vegetation; (b) flowering twig.

Erect herb with profuse branching to form dense vegetation, growing up to 0.65m high; leaves simple, alternate, sessile; lamina green or pink-green, round, ovate-oblong, 0.7–1.0 cm × 0.3–0.4 cm, isobilateral; inflorescence axillary spike; flower small, 0.1 cm × 0.15 cm; perianth: tepals 5, polytepalous, green; androecium: stamens 5, free; gynoecium: carpels 3, syncarpous; fruit drupe with persistent calyx, 0.10 cm × 0.15 cm.

Spot Identification
*Habit: erect herb with profuse branching.
**Leaf: green or pink-green.
***Inflorescence: axillary spike.

5.5.18 *UROCHONDRA SETULOSA* (TRIN.) C.E. HUBB. *(POACEAE)*

(Hooker's Icon. Pl. 35: t. 3457–1947.)

Description (Figure 5.114)

Grass up to 90 cm high; lamina convolute, 20.0–30.0 cm × 0.5–0.8 cm, ligule with a fringe of hairs, apex pungent; inflorescence panicle, 4.5–16.0 cm × 0.4–0.8 cm, spikelets, 0.2–0.3 cm, fertile spikelet sessile, oblong, laterally compressed; ovary beaked; fruit caryopsis, 0.2 cm × 3.5 cm, dark brown.

FIGURE 5.114 A twig with inflorescence.

Spot Identification
*Leaf: lamina convolute
**Inflorescence: panicle of spikelets

5.6 BACK MANGAL

Back mangals include 14 species arranged alphabetically, with each species described concisely.

5.6.1 CAESALPINIA BONDUC (L.) ROXB. (LEGUMINOSAE)

VN:B. – Nata; Arab. – Bunduk; Tam. – Kalarsikkodi; Tel. – Gacha; Mal. – Kalanji

Description (Figure 5.115)

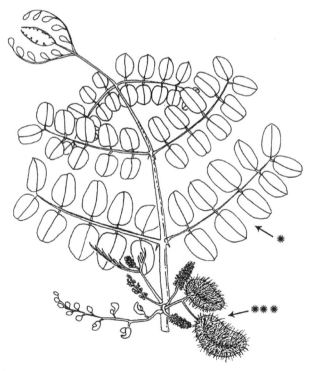

FIGURE 5.115   A twig with inflorescence.

Climbing shrub, diffuse branching; internodes with gregarious spines; leaves compound, secondary leaf rachis up to 25.0 cm long, alternate, paripinnate, petiolate; leaflet elliptic-oblong, 2.4–2.6 cm × 1.0–1.3 cm, mucronate; inflorescence raceme, lateral or axillary, up to 20.0 cm long; flower unisexual, irregular, with showy bracts; male flower 1.0 cm × 1.8 cm; calyx: sepal 5; corolla: petals 5, yellow; androecium: stamens 10, free; female flower 0.5 cm × 1.1 cm; gynoecium: carpels 2; fruit pod, 5.0 cm × 3.5 cm, flattened, gregarious spines round the surface.

Spot Identification
*Leaf: compound.
**Fruit: pod, flattened, gregarious spines round the surface.

5.6.2 *CAESALPINIA CRISTA* L. *(LEGUMINOSAE)*

VN: B. – Singri lata; S. – Kuberakshi; H.-Karanju; Bo. – Sagurghata; Ma. – Gajaga; Tam. – Kazhichikay; Tel. – Gachacha-kaya; Mal. – Kazanchik-kuru

Description (Figure 5.116)

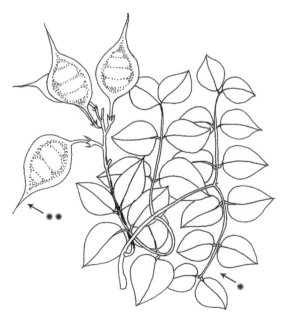

FIGURE 5.116   A twig with fruits.

Climbing shrub, diffuse branching to form thicket bush; internodes woody with frequent spines; leaves compound, alternate, paripinnate, leaflet rachis up to 25.0 cm long; leaflet elliptic-ovate, 3.0–3.5 cm × 1.2–1.5 cm, acute; inflorescence raceme, peduncle branched, up to 25.0 cm long; flower small; calyx: sepals 5, curved as boat shape; corolla: petals 5, dark yellow; androecium: stamens 10, base with dense white hairs; gynoecium: carpels 2, conical shaped; fruit pod with gradually pointed beak, 6.5 cm × 3.5 cm.

Spot Identification
*Leaf: compound.
**Fruit: pod, with gradually pointed beak.

5.6.3 *CYNOMETRA RAMIFLORA* L. *(LEGUMINOSAE)*

VN:B. – Shinger; S. – Madhuka; Mal. – Irripa; Tam. – Irudbu

Description (Figure 5.117)

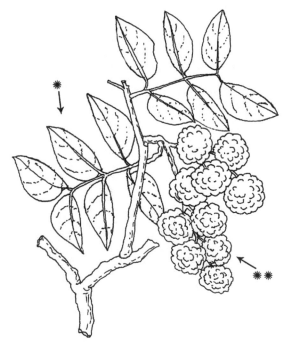

FIGURE 5.117   A twig with fruits.

Small to medium tree up to 15m high; leaves unijugate or bijugate, compound, paripinnate, leaflet rachis up to 8.0 cm long, lower pair of leaflets comparatively small; leaflet elliptic, 2.5–4.0 cm × 1.0–1.5 cm, obtuse, mid-vein obliquely run; inflorescence axillary raceme; flower 3.5 cm × 5.0 cm, reddish purple; calyx: sepals 4 lobed; corolla: petals 5, free; androecium: stamens 10, free; gynoecium: ovary inserted, stigma capitate; fruit elliptic-oblong, surface wrinkled, 2.5 cm.

Spot Identification
*Leaf: unijugate or bijugate, compound.
**Fruit: elliptic-oblong, wrinkled.

5.6.4   *DALBERGIA SPINOSA* ROXB. *(LEGUMINOSAE)*

VN:B. – Chulia-Kanta

Description (Figure 5.118)

FIGURE 5.118   A twig with inflorescence.

Shrub up to 6.0m high, profuse branching; leaves compound, impari-pinnate, alternate, stipulate, 2 woody spines at each node; leaflet with stalk, elliptic, 1.2 cm × 0.4 cm, obtuse; inflorescence raceme, panicles; flower zygomorphic; calyx: sepal 5; corolla: petals 5, white, vexillary; androecium: stamens 10 (9+1), diadelphous; gynoecium: carpels 2, stigma absent; fruit pod, kidney-shaped, flattened, 2.5 cm × 1.7 cm.

Spot Identification
*Flower: white, petals vexillary.

5.6.5   *DENDROPHTHOE FALCATA* (L.F.) ETTINGSH. *(LORANTHACEAE)*

(Denkschr. Kaiserl. Akad. Wiss., Wien. Math.-Naturwiss. Kl. 32: 52–1872.) VN:B. – Bara Manda; H.&P.-Banda; Guj. – Vando; S. – Vanda; Tam. – Pulluri; Tel. – Badanika

Description (Figure 5.119)

FIGURE 5.119   A twig with inflorescence.

Semi-parasitic or epiphytic plant growing on the mangrove tree (*Excoecaria agallocha*); normal roots absent but haustoria form to attach on the host trees; internodes lenticellate, woody; leaves simple, decussate, petiolate; lamina reddish, ovate-elliptic, 8.0–10.0 cm × 5.0–6.5 cm, obtuse; inflorescence raceme; flower up to 5.7 cm long; calyx: sepals 5, cup-shaped; corolla: petals 5, forming a hollow tube, showy, dark red; androecium: stamens 5, epipetalous; gynoecium: carpels 2, ovary syncarpous.

Spot Identification
*Habit: semi-parasitic or epiphytic plant.
**Flower: petals dark red.

5.6.6 *DERRIS SCANDENS* (ROXB.) BENTH. *(LEGUMINOSAE)*

VN: B. – Noa lata; H.-Gonj; P.-Gunj; Tam. – Takil; Tel. – Nalla tige

Description (Figure 5.120)

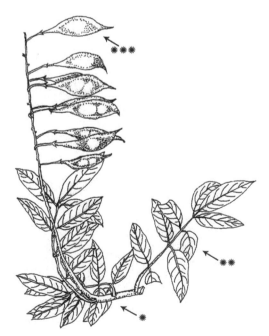

FIGURE 5.120 A twig with fruits.

Climbing shrub, profuse branching, occasionally twining; leaves compound, alternate, imparipinnate; leaflets more than 7 numbers, elliptic-lanceolate, 5.5–6.3 cm × 2.0–2.5 cm, entire; inflorescence raceme, peduncle up to 20.0 cm long; flower pedicellate, zygomorphic; calyx: sepals 5; corolla: petals 5, vexillary, white, stamens 10(9+1), diadelphous: gynoecium: carpels 2, stigma absent; fruit pod, 4.0 cm × 1.0 cm, shuttle shape, flattened, one margin thicker than other.

Spot Identification
*Habit: twining habit.
**Leaf: compound.
***Fruit: pod, shuttle shape.

5.6.7   *DERRIS TRIFOLIATA* LOUR. *(LEGUMINOSAE)*

VN: B. – Panlata; Bo. – Kirtana; Tel. – Tigekranuga

Description (Figures 5.121 and 5.122)

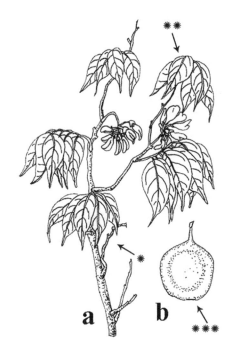

FIGURE 5.121   A twig with inflorescence.

FIGURE 5.122 A flowering twig.

Climbing shrub, profuse branching twining around the support; leaves compound, alternate, imparipinnate; leaflets 3–7, elliptic-oblong, 7.0–8.0 cm × 2.5–3.0 cm, entire; inflorescence raceme, peduncle up to 8.0 cm long; flower pedicellate, zygomorphic, 1.0 cm long; calyx: sepals 5; corolla: petals 5, vexillary, white-pinkish; androecium: stamens 10(9+1), diadelphous; gynoecium: carpels 2, stigma absent; fruit pod, 3.2 cm × 2.5 cm, sub-orbicular.

Spot Identification
*Habit: twining habit.
**Leaf: leaflet 3–7.
***Fruit: pod, sub-orbicular.

5.6.8 *FLAGELLARIA INDICA* L. *(FLAGELLARIACEAE)*

(Sp. Pl. 333–1753.)

VN: Tam. & Mal. – Paanambuvalli; Tel. – Vanachandra;

Description (Figure 5.123)

Climbing shrub with sparse branching up to 15m high; cane-like stem; leaves simple, alternate, petiole with distinct sheath covering internode; lamina lanceolate, 10.0–40.0 cm × 5.0–20.0 cm, coiled apex forms tendril, base auriculate; inflorescence panicle, profuse branching, up to 25.0 cm long; flowers clustered; perianth: tepals white, ovate; fruit drupe, up to 0.5 cm, globose.

FIGURE 5.123   A flowering twig.

Spot Identification

*Habit: thick cane-like stem.

**Leaf: coiled apex forming tendril.

***Inflorescence: panicle, profuse branching.

5.6.9 *MEROPE ANGULATA* SWINGLE *(RUTACEAE)*

(J. Wash. Acad. Sci, 5; 423–1915,), *an unresolved name as per the web (www.theplantlist.org).
VN: Ban lebu

Description (Figures 5.124 and 5.125)

Shrub up to 3.0m long; bark blackish, fissured; leaves simple, alternate, stipules as two spines at each node, petiolate; lamina oblong-lanceolate, 5.0–7.0 cm × 3.0–4.0 cm, lemon aroma, acute; inflorescence solitary; flower, 1.8 cm × 0.5 cm; calyx: sepals 5; corolla: petals 5, white, fleshy; androecium: stamens 10; gynoecium: carpel 1: fruit berry, with triangular ridges, green.

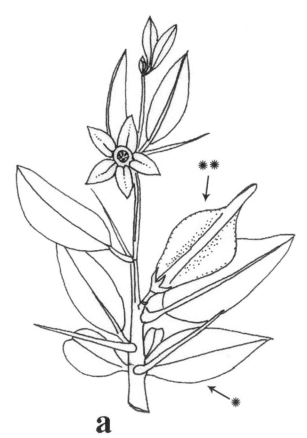

FIGURE 5.124 A twig with flower & fruit.

FIGURE 5.125   A twig with fruit.

Spot Identification
*Leaf: lemon aroma.
**Fruit: berry with triangular ridges.

5.6.10 *PANDANUS FURCATUS* ROXB. *(PANDANACEAE)*

(Hort. Bengal. 71–1814.)

VN:B. – Keya; S. – Ketaka; H. Keora; Bo. – Keura; Tam. – Talhai; Tel. – Ketaki; Mal. – Tala

Description (Figures 5.126 and 5.127)

Small tree up to 5.0m high, dichotomously branched; stem palm like structure supported by aerial roots; leaf sword shape, up to 2.0 long, serrated/dented, acuminate; inflorescence spadix; flower strong aromatic; gynoecium: carpels confluents, stigma short; fruit oblong, drupes aggregate.

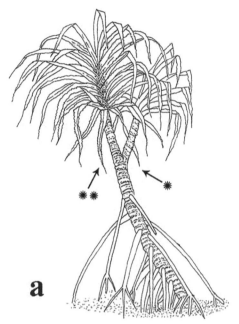

FIGURE 5.126   A plant with stilt roots.

FIGURE 5.127   Plant with fruits.

Spot Identification
*Habit: dichotomously branched.
**Leaf: sword shape, serrated.

5.6.11 *PONGAMIA PINNATA* (L.) PIERRE *(LEGUMINOSAE)*

RLC&C: LC (ver 3.1); YP: 2012; DA: 2010–07–27
VN: B., S. &H.-Karanja; Bo.&P.-Karanj; Mal. – Punnu; Tam. – Pungu;
Tel. – Kranuga

Description (Figure 5.128)

   Medium tree up to 25.0m high; bark brownish black; leaves compound, alternate, imparipinnate; leaflet stalked, ovate-lanceolate, 8.0–10.0 cm × 4.0–5.0 cm, acute; inflorescence raceme; flower zygomorphic, up to 1.2 cm long; calyx: sepals 5; corolla: petals 5, vexillary, white-pinkish; androecium: stamens 10 (9+1), diadelphous; gynoecium: carpels 2; fruit pod, woody, 5.8 cm × 3.5 cm; seed nut-like, black brown on maturity.

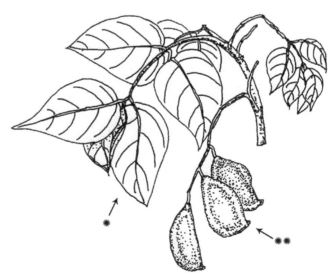

FIGURE 5.128   A twig with fruits.

Spot Identification
*Leaf: compound.
**Fruit: Pod, woody.

5.6.12 *SOLANUM TRILOBATUM L. (SOLANACEAE)

*An unresolved name as per the web (www.theplantlist.org).
VN:B. – Lata Gurbegun; S. – Alarka; Kan. – Mullumusta; Mal. – Tutav-alam; Tam. – Tutuvalai; Tel. – Tel-lavuste

Description (Figure 5.129)

Trailing shrub with hooked spines; internode covered with flattened spines; leaves simple, alternate, petiolate; lamina trilobate, 2.5–3.0 cm × 1.5–2.0 cm, dorsal surface light violet, ventral surface green; inflorescence corymbose raceme; flower 2.5 cm × 1.5 cm; calyx: sepals 5; corolla: petals 5, violet, bell shape; androecium: stamens 5, epipetalous; gynoecium: carpels 2, ovary syncarpous; fruit berry, up to 1.0 cm long, globose with persistent calyx, deep red at maturity.

FIGURE 5.129 A twig with flowers and fruits.

Spot Identification
*Habit: trailing shrub.
**Leaf: lamina trilobite.
***Flower: petals violet.

5.6.13   *TAMARIX GALLICA* L. *(TAMARICACEAE)*

(Sp. Pl. 270–1753.)
RLC&C: LC (ver 3.1); YP: 2014; DA: 2013–04–25
VN:B. – Nona jhau; S. – Jhavuka; H. – Jhau; Bo. & Guj. – Javnu-jhadu;
Mal. – Jhavukam; Tam. Sirusavukku; Tel. – Sirasaru; P. – Pilchi

Description (Figure 5.130)

FIGURE 5.130   (a) A twig with inflorescences; (b) leaf.

Shrub with deliquiscent branching up to 10.0 m high; bark black brown, longitudinally fissured, leaf scars in internodes; leaves simple, decussate, sessile; lamina triangular, 0.15 cm × 0.15 cm, tapering end; inflorescence catkin up to 12.0 cm long, peduncle branched; flower small, 0.1 cm × 0.05 cm; calyx: sepals 5; corolla: petals 5, pink-red, persistent; androecium: stamens 5, free; gynoecium: carpels 3, stigma short, globose; fruit capsule with persistent calyx, red violet.

Spot Identification
*Leaf: lamina triangular, 0.15 cm × 0.15 cm.
**Inflorescence: catkin.

5.6.14   *VISCUM ORIENTALE WILLD. (LORANTHACEAE)

*Considered as a synonym of *Viscum cruciatum* Sieber ex Boiss. (Santalaceae) *Voy. Bot. Espagne 2: 274–1840.* as per the web (www.theplantlist.org) VN:B. – Manda; Tel. – Sundarabadanika

Description (Figure 5.131)

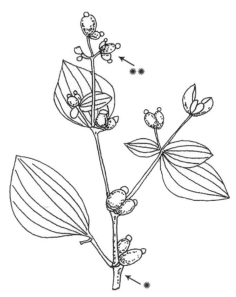

FIGURE 5.131   A twig with inflorescences.

Semi-parasitic or epiphytic, evergreen plant mainly growing on the branches of *Excoecaria agallocha* and *Xylocarpus* spp. trees; root forming haustorium to anchor the branch of the host tree; stem with profuse branching, nodes ridged, easily dissociated, flattened; leaves simple, decussate, petiole continued to lamina; lamina various shape: elliptic-lanceolate, lanceolate, reniform, 4.0–6.0 cm × 0.8–2.0 cm; multicostate, parallel convergent venation; inflorescence cyme, cup-shaped, capitulum head, green; flower small up to 0.1 cm; perianth: tepals 4, polytepalous, green, fleshy; gynoecium: stigma globose;fruit green, fleshy.

Spot Identification
*Leaf: Semi-parasitic or epiphytic plant.
**Flower: tepals 4, green, fleshy.

(Threatened categories and vernacular names of the intertidal flora are mentioned following IUCN Red List of Threatened species, 2008 and Chopra et al., 1986, respectively.)

KEYWORDS

- back mangals
- description
- major mangroves
- mangrove associates
- mangrove halophytes
- salient morphometry
- spot identification

FIGURE 5.2   (a) Canopy; (b) a flowering twig; (c) fruits.

FIGURE 5.4   (a) Canopy; (b) a flowering twig; (c) fruits.

FIGURE 5.6   (a) Canopy; (b) leaf; (c) flower; (d) a twig with fruits.

FIGURE 5.8   (a) Canopy; (b) flowering buds; (c) flowers; (d) fruits.

FIGURE 5.10   (a) Canopy; (b) leaf; (c) flowering twig; (d) fruits.

FIGURE 5.12   (a) Canopy; (b) leaf; (c) flowering twig; (d) fruit.

FIGURE 5.14   (a) Canopy; (b) leaf; (c) flowering twig; (d) fruits.

FIGURE 5.16 (a) Canopy; (b) flowering twig; (c) hypocotyls hanging from mother tree; (d) hypocotyl.

FIGURE 5.18 (a) Canopy; (b) knee roots; (c) young tree; (d) hypocotyls hanging from mother tree.

FIGURE 5.20   (a) Canopy; *b*. knee roots; (c) flowering twig; (d) hypocotyls hanging from mother tree.

FIGURE 5.22   (a) Canopy; (b) flowering twig; (c) hypocotyl.

FIGURE 5.24 (a) Canopy; (b) flowering twig; (c) hypocotyls attached with mother tree.

FIGURE 5.26 (a) Canopy; (b) flowering twig; (c) hypocotyls hanging from mother tree.

FIGURE 5.28   (a) Canopy of male tree; (b) root; (c) male inflorescence; (d) female inflorescence.

FIGURE 5.30   (a) Flowering twigs; (b) leaf; (c) fruiting branch.

FIGURE 5.32   (a) Canopy; (b) inflorescence; (c) a twig with fruits; (d) fruit.

FIGURE 5.34   (a) Canopy; (b) a bunch of hypocotyls hanging from mother tree; (c) hypocotyl.

FIGURE 5.36   (a) Canopy; (b) flowering twig; (c) fruits.

FIGURE 5.38   (a) Canopy; (b) flowering twig; (c) flowers.

FIGURE 5.40   (a) Canopy; (b) male inflorescence; (c) female inflorescence; (d) fruits with incipient vivipary.

FIGURE 5.42 (a) Canopy; (b) twig with fruits.

FIGURE 5.44   (a) Canopy; (b) twig with flower buds; (c) twig with hypocotyls.

FIGURE 5.46 (a) Canopy; (b) twig with flower buds; (c) twig with hypocotyls.

FIGURE 5.48   (a) Canopy; (b) flower showing distinct style.

FIGURE 5.50   (a) Canopy; (b) inflorescence; (c) fruits.

FIGURE 5.52   (a) Canopy; (b) flowering twig.

FIGURE 5.54   (a) Canopy; (b) flower; (c) a bunch of fruits.

FIGURE 5.56   (a) Canopy; (b) flowering bud; (c) fruit.

FIGURE 5.58   (a) Canopy; (b) flowering bud; (c) fruits.

FIGURE 5.60   (a) Canopy; (b) pneumatophores; (c) leaf; (d) flower; (e) fruit.

FIGURE 5.62   (a) Canopy; (b) inflorescence; (c) twig with fruit.

FIGURE 5.64   (a) Canopy; (b) inflorescence; (c) fruits.

CLIMATE CHANGE RESPONSE AND MANAGEMENT OF MANGROVES

6.1 CLIMATE CHANGE: AN OVERVIEW

Climate change is a long-term shift in weather conditions identified as changes in temperature, precipitation, moisture in winds, wind pressure and other atmospheric indicators. Climate change occurs due to both changes in average conditions and changes in variability, including extreme events. The earth's climate is naturally variable on a geological timescale. However, its long-term state and average temperature are regulated by the balance between incoming solar radiation and outgoing earth's radiation, which determines the earth's energy balance or heat budget. Any factor that causes a sustained change to the amount of incoming energy to the earth or the amount of outgoing energy from the earth can lead to climate change. As these factors are external to the climate system, they are referred to as 'climate forcers,' invoking the idea that they force or push the climate towards a new long-term state – either warmer or cooler depending on the direction of change. The causes of climate change can be divided into two categories – those that occur due to natural causes and those that are created by humans. What the world is more worried about is that the changes that are occurring today have been speeded up because of different human activities.

There are a number of natural factors responsible for climate change. Some of the more prominent ones are drifting of continents, volcanic eruption, ocean currents, the earth's tilting position to its axis, including temporal and spatial variations in the sun's energy reaching earth, etc. These factors have been responsible for earth's climate change geologically several times. Scientists have collated a number of facts which have been evident to be a record of earth's climate change, dating back hundreds of thousands of years (and, in some cases, millions or hundreds of millions of years). The evidences which have been analyzed through a number of indirect measures in relation to climate change include tree rings, pollen remains, ice cores, glacier lengths, and ocean sediments. Scientists also studied changes in earth's orbit around the sun (IPCC, 2013).

Around 1750 the Industrial Revolution started; climate change occurred rapidly due to human activities that have contributed substantially by adding CO_2 and other heat-trapping gases to the atmosphere. These gases are called greenhouse gases (GHGs). GHGs have augmented the effect of greenhouse and thereby facilitated temperature to rise in the earth's surface. Initially, human activities have been responsible for emission of greenhouse gas mostly due to burning of fossil fuels followed by the conversion of forest land for rehabilitation and agriculture. The cumulative effect of all the phenomena is related to the amount and rate of climate change in severity.

The Kyoto Protocol (1997) identified six gases as major GHGs. They are carbon-dioxide (CO_2), methane (CH_4), nitrous oxide (N_2O), hydrofluorocarbons (HFCs), perfluorocarbons (PFCs), and sulfur hexafluoride (SF_6). According to Kiehl and Trenberth (1997), the Greenhouse 'effectiveness' on a clear day is: 60% water vapor; 25% carbon dioxide; 8% ozone and 7% trace gases, for example, nitrous oxide, methane. Most studies dealing with climate change, however, have almost exclusively focused on CO_2, which remains the main anthropogenic contributor to climate change. Importantly, CO_2 remains longer in the atmosphere than the other major GHGs. Consequently, over the last century the global climate has warmed by an average of 0.6°C as much as seen over the last 30 years. The IPCC has predicted that the rise in temperature will continue leading to a rise in the range of 1.4°C to almost 6°C by 2100 A.D. At the same time, global average sea levels are also predicted to rise by 9 to 88 cm by 2100 A.D. (Panda and Mishra, 2017).

6.1.1 CHANGING ECOSYSTEM IN RELATION TO CLIMATE CHANGE

Climate change is expected to cause an irreversible change and alteration in the ecosystems. Climate change may directly affect ecosystems by changing or threatening biodiversity of both plant and animal communities in any particular location. Although species have adapted to environmental change for millions of years, a rapidly changing climate of today could require adaptation on larger and faster scales than in the past. Those species that cannot adapt are at risk of survival. In the past several lines of evidence were there when climate change caused widespread species extinction, migration and behavior changes. Warming-induced shifts in species composition have been broadly observed in grasslands (Yang *et al.*, 2011) and tundra ecosystems (Walker *et al.*, 2006). Many attributes of organisms at the individual, population and community levels, which are related to biological and ecological changes, include tolerance to physicochemical factors, the ability to compete one another for limiting resources, and persistence to functional processes such as ingestion, growth, and respiration rates (Alongi, 2008). Organisms in all ecosystems are also subject to encounter a variety of disturbances arising due to both natural and anthropogenic activities. All the activities vary in their duration, frequency, size, and intensity that play an important role in adaptive change (Odum and Barrett, 2004).

6.1.2   THE IMPACT OF CLIMATE CHANGE ON MANGROVES FORESTS

Significant components of climate include temperature, atmospheric moisture including cloud cover and precipitation, pressure and wind system. However, atmospheric CO_2 concentration, considered one of the GHGs, does not come directly under climatic component, though it is considered one of the important factors in changing climate. Change of each component affects mangroves and cumulative effect of all components may cause a deterioration of mangroves in a particular environmental setting. Mangroves respond to a variety of climatic changes of geological, physical, chemical and biological phenomena with temporal and spatial patterns. Mangroves that occupy the interface between land and sea at low latitudes usually face a harsh environment as they are inundated twice a day, and affected by fluctuation of tidal changes, water and salt exposure, and varying degrees of anoxia due to submergence (Tomlinson, 1986; Naskar and Mandal, 1999; Saenger, 2002; Alongi, 2008; Naskar and Palit, 2015). Mangroves are, therefore, highly tolerant to waterlogged saline soil and exhibit a high degree of ecological stability (Saenger, 2002; Alongi, 2008; Naskar and Palit, 2015).

However, among all the outcomes from climate changes, relative sea-level rise may be considered to be the greatest threat for alterations to land surfaces (Field, 1995; Lovelock and Ellison, 2007). It may reduce mangroves and other tidal wetlands areas and affect the health of mangroves in recent and predicted future (IUCN, 1989; Ellison and Stoddart, 1991; Nichols et al., 1999; Ellison, 2000; Cahoon and Hensel, 2006; McLeod and Salm, 2006; Gilman et al., 2006, 2007a, b; Ward et al., 2016; Wilson, 2017). In contrast until now, anthropogenic activities for conversion of mangroves areas into agriculture, aquaculture, fisheries, along with resources collection have been a major threat to mangroves ecosystems (IUCN, 1989; Primavera, 1997; Valiela et al., 2001; Alongi, 2002; Duke et al., 2007; Das and Mandal, 2016).

6.2 IMPORTANCE OF MANGROVE ECOSYSTEMS

Mangrove ecosystem services which have staggering benefits to human welfare, coastal people in particular, are not accounted properly (Mandal et al. 2010). These tidal forests are important sites of breeding and nursery grounds for fish, shellfish, crustaceans, reptiles, birds and mammals; are repositories of renewable resource like wood and other forest products; and act as sink for accumulation of sediment, nutrients, and contaminants (Twilley, 1995; Kathiresan and Bingham, 2001; Manson et al., 2005). Mangroves render services such as protection from strong winds, tidal thrust, sea surges, tsunamis, and shoreline erosion (Mazda et al., 2007; Godoy and Lacerda, 2015). And their ecosystems are considered to be the most important sources of a variety of coastal bio-resources and thus support sustainability of coastal people through livelihoods, particularly in developing countries (Alongi, 2002), who are traditionally dependent on mangrove resources such as food, fodder, timber, fuel, medicine, honey, wax and other products (Saenger, 2002; Das and Mandal, 2016). Mangrove ecosystems are really comprised of two systems such as terrestrial and aquatic; two systems are considered to be integrated and complementary one another, enriched with resources of high economic values.

6.2.1 AQUATIC RESOURCES

Aquatic systems surrounding mangroves are rich in fishery resources. Many fishery species are permanent residents: some spending their entire life cycle, some have temporary long-term residents, few are associated with mangroves during at least one stage of life cycle, and others have sporadic short-term residents or sporadic users of mangroves (Robertson and Duke, 1990; Ogden, 1997; Barletta-Bergan et al., 2002; Nagelkerken et al., 2002; Crona and Ronnback, 2007; Serafy and Araujo, 2007; Das and Mandal, 2016). A variety of mud crabs for which mangroves water bodies are ideal habitats have been the most lucrative ones of all resources in terms of market demand as well as price. Mangroves aquatic systems act as breeding and nursery grounds for early life stages of several aquatic fauna including mud crab (Matthes and Kapetsky, 1988; Bagarinao, 1994; Primavera, 1998; Walton et al., 2006; Cannicci et al., 2008; Nagelkerken et al., 2008). Changes in sea levels are the most obvious threat to mangrove aquatic ecosystems resulting into gradually declining aquatic resources worldwide.

6.2.2 TERRESTRIAL RESOURCES

A biological product that is harvested from a forested area is commonly termed a 'non-timber forest product (NTFP) (Shackleton and Shackleton, 2004). Non-timber forest products (NTFP) are also recognized as important economic sources for coastal people worldwide (Vedeld et al., 2004). Honey, Bee wax, wood, *Nypa fruticans, Phoenix paludosa*, etc. are the non-timber mangrove forest products other than aquatic resources which are usually collected by the villagers living adjoining forest areas. Honey and pollen, treated as NWFP (Non wood forest's products), are used as traditional medicine, high-energy food, and source of vitamins and minerals (Saenger, 1983; Hamilton and Snedekar, 1984; Untawale, 1987; Adegbehin, 1993; FAO, 1994; Basit, 1995; Sathirathai and Barbier, 2001; Nagelkerken et al., 2008; Das and Mandal, 2016).

Timber forest product includes wood which has two distinct uses: domestic fuel and construction (FAO, 1994; Bandaranayake, 1998). Coastal people as traditional practice use mangrove woods for fuel. The quality of mangrove woods burning long has been reported for its commercial utility in bakeries and clay firing kilns (Lacerda et al., 1993;

Naylor et al., 2002; Walters, 2004). They use mangrove wood for a variety of purposes, including constructions such as posts, beam, roofing, fencing, thatching with 'nipa' palm, *Nypa fruticans*. They also make various craft items such as fish traps/wires (Adegbehin, 1993; Rasolofo, 1997; Ewel et al., 1998; Semesi, 1998; Kovacs, 1999; Primavera et al., 2004; Walters, 2004). A great many people rely on these products to meet subsistence needs as an important income supplement (Christensen, 1982; FAO, 1985; 1994; Kunstadter et al., 1986; Diop, 1993; Lacerda et al., 1993; Spalding et al., 1997; Glaser, 2003; Walters, 2005; Lopez-Hoffman et al., 2006; Rönnbäck et al., 2007; Das and Mandal, 2016).

6.3   EFFECTS OF MANGROVES LOSS

Dwindling mangrove areas due to climate change cause irreparable damage to shorelines and increase the threat to safety of coastal inhabitants. The threats that coastal area encounter due to mangrove loss are of two types in ecological point of view: (i) abiotic and (ii) biotic disturbances.

i)   Abiotic disturbances like flooding, storm waves, cyclones, sea surges, tsunami, etc. may cause immediate effects with high degree of intensification, such as soil erosion, soil salinity, soil pH, saline soil ingression, loss of land productivity, etc. These are very common and recently observed phenomena worldwide (Naskar and Mandal, 1999; Danielsen et al., 2005; Kathiresan and Rajendran, 2005; Dahdouh-Guebas et al., 2005a, b, 2006; Winterwarp et al., 2013; Lang'at et al., 2014). The above effects may alter the overall environmental setting of coastal habitats, and thereby spoil coastal water quality, release a colossal amount of stored carbon, aggravate global warming and affect other climate change trends (Ramsar Secretariat, 2001; Kristensen et al., 2008).

ii) Biotic disturbances may affect mangrove losses which indirectly cause to eliminate diverse groups of fishery species including fishes, crustaceans, crabs and molluscs as their nursery beds are either greatly disturbed or poorly enriched from scarcity of nutrients supply.

Biotic disturbances may directly affect diverse gene pools of mangrove regions. So, mangroves ecosystems of high biological diversity on which coastal people rely on their livelihood will be affected to a great extent (Ewel et al., 1998; Mumby et al., 2004; Nagelkerken et al., 2008; Walters et al., 2008; Das and Mandal 2016). Evidently mangrove regions, altogether, provide versatile services to both the people and environment. The annual economic values of mangroves, including the products cost and services they provide, have been estimated to be US$ 194,000–900,000/ha/year (Wells et al., 2006; Brander et al., 2012; Costanza et al., 2014).

6.4 THREATS TO MANGROVES FROM CLIMATE CHANGE

The climatic components which are intensified over the times, primarily due to anthropogenic activities, appear to be a threat to overall Earth's ecosystems.

6.4.1 TEMPERATURE

The earth has warmed 0.6–0.8°C since 1880, and is projected to warm 2–6°C by 2100 mostly due to anthropogenic activity (Houghton et al. 2001). Field (1995) opined that as expected mangroves are unlikely to be affected extremely by the projected increases in sea temperature. Effect of temperature facilitates mangroves to produce optimal shoot density at 25°C as mean air temperature and stop sprouting leaves below 15°C (Hutchings and Saenger, 1987). Some mangroves exhibit a declining trend of leaf formation rate above 25°C, (Saenger and Moverly, 1985) and also have thermal stress above 35°C that affects root structures and seedlings establishment of mangroves (UNESCO, 1992). The optimal temperature required for photosynthesis of mangroves varies between 28°C and 32°C; however, no photosynthesis occurs in leaves when temperature rises above 38–40°C (Clough et al., 1982; Andrews et al., 1984).

There are possibilities for mangroves to migrate to higher latitudes due to temperature rise as predicted by some scientists (UNEP, 1994; Field, 1995; Ellison, 2005). However, their migration polewards may be restricted by extreme cold events (Woodroffe and Grindrod, 1991) since temperature is likely to affect mangroves with some physiological events: (i) photosynthesis, (ii) water loss, (iii) transpiration, and (iv) salt loss (Pernetta, 1993).

Furthermore, if mangroves are affected by increasing surface temperature, then following events may occur (Field, 1995; Ellison, 2000) as;

i) Species composition may change.
ii) Phenological patterns, including timing of flowering and fruiting, may vary.
iii) Photosynthesis may occur normally.
iv) Productivity may increase where temperature does not exceed an upper threshold.
v) Supply of propagules may be certain.

Edwards (1995) argued that despite the uncertainties of temperature variation that may affect the species composition, vegetative structure or the seasonal patterns of reproduction and flowering of mangroves, an increase in sea-surface and air temperatures is likely to benefit mangroves living near the poleward limits of present distributions. The increasing temperature not only facilitates mangroves to migrate polewards, but also favors the events such as:

i) Increasing species diversity.
ii) Increasing litter production.
iii) Promoting growth of larger trees.
iv) Favoring changes of seasonal patterns of reproduction.
v) Increasing the length of time between flowering and the fall of mature propagules (UNEP, 1994; Ellison, 2000).
vi) Increasing growth rates of bacteria due to increased sediment temperature may result in enhanced rates of nutrient recycling and mangroves regeneration (UNEP, 1994).

Temperature changes may substantially affect tolerance of mangroves, which greatly vary from species to species. Li and Lee (1997) grouped the mangrove species in China into three classes based on their thermal tolerance:

(a) Cold-resistant eurytopic species (e.g., *Kandelia candel*, *Avicennia marina* and *Aegiceras corniculatum*).
(b) Cold-intolerant (thermophilic) stenotopic species (e.g., *Rhizophora mucronata*, *R. apiculata*, *Lumnitzera littorea*, *Nypa fruticans* and *Pemphis acidula*).

(c) Thermophilic eurytopic species, (e.g., *R. stylosa, Bruguiera sexangula, B. gymnorhiza, Excoecaria agallocha* and *Acrostichum aureum* (Zhang and Lin, 1984).

6.4.2 ATMOSPHERIC CO_2 CONCENTRATION

Atmospheric CO_2 has increased from 280 parts per million by volume (ppmv) in the year 1880 to nearly 400 ppmv in the year 2012. This increase of CO_2 concentration is a 35% increase from a pre-industrial value (1757AD). In recent decades, CO_2 emission is projected to continue to increase. One of the views that most atmospheric CO_2 increase that results from burning of fossil fuels will be absorbed into the ocean, affecting ocean chemistry (McLeod and Salm, 2006).

Increased levels of CO_2 are expected to enhance photosynthesis and mangrove growth rates, with effective change of vegetation structure along natural salinity and aridity gradients (Gilman et al. 2008). Ball et al. (1997) mentioned that increased CO_2 did not change growth rates of mangroves in high salinity, but facilitated mangroves to have faster growth with increasing biomass and branching activity in low salinity, subject to availability of freshwater (Farnsworth et al., 1996). Flowering period may occur earlier than normal, and pollinators may desynchronize with plants (Alongi, 2008). Advanced mangroves under these conditions showed earlier maturation than ambient controls. However, when enhanced CO_2 is combined with reduced rainfall, this condition creates stress that affects mangrove growth (Gilman et al., 2008). Elevated CO_2 may promote the mangroves growth when evaporative demand restricts carbon gain at the leaves.

In arid regions, mangroves may benefit due to the increase in water use efficiency, because decreased water loss through transpiration will accompany CO_2 uptake (Ball and Munns, 1992). However, mangroves may lose such an advantage if salinity increases in arid regions. CO_2 increase is unlikely to affect mangrove growth when salinity remain too high for a species to maintain water uptake. Global CO_2 increases may lead mangroves to have a competitive advantage in arid regions, with their ability to reduce water use in water stress condition and to maintain relatively high rates of CO_2 uptake (Snedaker and Araújo, 1998).

Not that all mangroves species will respond similarly to increasing CO_2, but that species-specific responses will vary to the interactive effects of rising CO_2. Species patterns are greatly influenced within estuaries under specific environmental conditions including temperature, salinity, sea level change, nutrient levels, and hydrologic regime; cumulative effect of all the factors influences how a mangrove responds to increasing CO_2 (Field, 1995).

6.4.3 PRECIPITATION

Globally, rainfall is projected to increase by about 25% by 2050 due to global warming. However, the distribution of rainfall will be uneven from region to region (Knutson and Tuleya, 1999; Walsh and Ryan, 2000; Houghton et al., 2001). Precipitation may increase in high-latitudes, and decrease in most subtropical regions, especially at the poleward margins of the subtropics (Solomon et al., 2007). Intergovernmental Panel on Climate Change (IPCC) reported that precipitation would increases significantly in eastern parts of North and South America, northern Europe and northern and central Asia. On the other hand, it reported that precipitation would decrease significantly in the Sahel, the Mediterranean, southern Africa and parts of southern Asia (Solomon et al., 2007).

Changes in precipitation patterns due to climate change may affect adversely both mangroves growth and their expansion (Field, 1995; Snedaker, 1995). However, it is not clear to predict what will happen to mangroves under the conditions of decreased and increased precipitations, based primarily on links observed between mangrove habitat condition and rainfall trends (Field, 1995; Duke et al., 1998)?

The following events may occur when rainfall reduces.

i)   Decreased precipitation and increased evaporation will enhance salinity, leading to decrease in net primary productivity, growth and seedling survival. Decreased precipitation may lead to competition among mangrove species, alteration in species composition, decreasing the diversity of mangrove zones, causing a projected loss of the landward zone to be turned into high salt affected flats without vegetation (Snedaker, 1995; Ellison, 2000, 2004).

ii) Decreased precipitation would lead to deficit amount of freshwater that may result in less surface water input to mangroves and lesser water input into the groundwater which, in turn, increases salinity. Increasing salinity in the soil leads to deposit higher amount of salt into mangrove tissues. The higher amount of salinity and lesser amount of freshwater affect mangrove productivity, growth, and seedling survival, and may alter species composition in one way and may favor more salt-tolerant mangroves in other (Field, 1995; Ellison, 2000, 2004). Apart from that, increased salinity will enhance sulfate amount in seawater that would facilitate anaerobic decomposition of peat, which altogether would make mangroves vulnerable to any rise in relative sea-level (Snedaker, 1993, 1995).

iii) Decreased precipitation may lead mangroves to encroach into salt marsh and freshwater wetlands (Saintilan and Wilton, 2001; Rogers et al., 2005a).

The following events may occur when rainfall increases.

i) Increased rainfall may enhance growth rates of mangroves, resulting in an increase of mangroves area and diversity of mangrove zones, with the occupation of earlier unvegetated areas of the landward fringe within the tidal wetland zone (Field, 1995; Duke et al., 1998). Mangroves may also migrate and outcompete other vegetation in the increased precipitation (Harty, 2004).

ii) Increased precipitation may cause higher diversity of mangroves and their productivity. Probably, this would occur due to sufficient amount of fluvial sediment and nutrients, and the least exposure of mangroves to sulfate and reduced salinity (McKee, 1993; Field, 1995; Ellison, 2000). Mangroves are likely to increase peat production with increased freshwater inputs and concomitantly reduced salinity due to decreased sulfate exposure (Snedaker, 1993, 1995).

iii) Increased precipitation may cause decreasing salinity that facilitates growth rates to increase in certain mangroves species (Field, 1995), because growth of mangroves has been linked to low salinities (Burchett et al., 1984; Clough, 1984).

6.4.4   STORMS AND HURRICANES

International Panel on Climate Change (IPCC, 2007) predicted that wind intensities is likely to increase by 5 to 10 percent (Houghton et al., 2001), although no recorded trends have been observed in tropical storms, with no evidence of areas in storm formation or its frequency changes. However, a more recent assessment predicts that tropical storms will increase in both frequency and intensity because of climate change (Trenberth, 2005) and would disturb mangroves and its ecosystems. It also leads ocean circulation to change that is likely to influence regional rates of sea-level rise (Stammer et al., 2013).

According to IPCC during the twenty-first century, global warming may cause tropical cyclone peak wind intensities to increase and enhancement of precipitation intensities of cyclone mean and peak in some areas (Houghton et al., 2001; Solomon et al., 2007; Woodroff et al., 2013). Combination of both increase in strong winds frequency and low pressures causes enhancement of storm surges height, when storms are more frequent or severe due to climate change (Church et al., 2001; Houghton et al., 2001; Solomon et al., 2007). In combination with hurricanes, storms with increased intensity and frequency may disturb mangroves through alteration of its ecological setting as:

- **Alteration of mangroves environmental setting:** Storms can alter mangrove sediment elevation and budget through soil deposition, soil erosion, soil compression, soil salinity, flood inundation, and peat collapse (Smith et al.,1994;Woodroffe and Grime, 1999; Baldwin et al., 2001; Church et al., 2001, 2004a, 2004b; Sherman et al., 2001; Woodroffe, 2002; Cahoon et al., 2003, 2006; Cahoon and Hensel, 2006; Piou et al., 2006; Gilman et al., 2006). Hurricanes can uproot trees, leaving sediment unprotected and vulnerable to erosion and thereby aggravate the impacts of sea-level rise. Also, increased levels and high frequency of extremely high water may alter sediment elevation and sulfide soil toxicity; however, the knowledge on ecosystem effects due to changes by extreme waters is insufficient (Gilman et al., 2008).
- **Alteration of structure of mangroves:** It is expected that increase in hurricane intensity over the next century is likely to decrease the average height of mangroves (Ning et al., 2003), because increased intensity and frequency of storms would have strong impact to damage severely mangroves through defoliation, tree mortality, stress, and sulfide soil toxicity, and thus lead to a change in community structure.
- **Alteration of mangroves ecosystems:** Areas suffering from mass tree mortality with little survival of saplings and trees may also

lead to permanent ecosystem alteration. Restoration of ecosystem may not be possible, because seedlings recruitment is unlikely to establish due to alteration in sediment elevation and concomitant changes in hydrology (Cahoon et al., 2003; Cahoon, 2006). Roth (1997) opines that high storm may cause to shift species proportions because respective mangroves have different rates of regeneration. Other natural adversities, such as tsunami can also devastate mangroves and disturb other coastal ecosystems to a great extent, for example, the 26 December 2004 Indian Ocean tsunami (Danielsen et al., 2005; Kathiresan and Rajendran, 2005; Dahdouh-Guebas et al., 2005a, b, 2006).

- **Alteration of function of mangroves:** Storm surges can also inundate mangroves in combination with sea-level rise. This may affect mangroves photosynthesis, productivity, and survival (Ellison, 2000). Submergence of lenticels in the aerial roots due to inundation can decrease oxygen concentrations in the mangroves, resulting in mass mortality (Ellison, 2004). Inundation is also projected to reduce the ability of mangrove leaves to conduct water and to photosynthesize (Naidoo, 1983).

6.4.5 OCEAN CIRCULATION PATTERNS

Covering about 70.9% of earth's surface and absorbing about twice as much of the sun's radiation as the atmosphere or the land surface, the oceans play a more active and dynamic role in the climate system. Ocean currents, either hot or cold, move as vast amounts of heat generated from one region to another. But, here, heat transfer is more localized and channeled into specific regions. IPCC reports that presently ocean circulation change seems stable as before (Bindoff et al., 2007). However, changes of long-term trends in global and basin-scale ocean heat content may occur with salinity increase, which are related to changes in ocean circulation (Gregory et al., 2005; Bindoff et al., 2007). Ocean surface circulation patterns changes may affect dispersal of mangrove propagules and the genetic pool of mangroves diversity, with concomitant effects on structure of mangrove community (Duke et al., 1998; Benzie, 1999; Lovelock and Ellison, 2007). Increasing gene flow between recently separated populations and increasing diversity of mangrove species may facilitate mangroves responses to be much resistant and resilient to climate change (Gilman et al., 2008).

6.5   SEA-LEVEL RISE (SLR)

Global warming causes sea-level rise, the greatest challenge of climate change. Sea-level rise is predicted to be the major threat to certain ecosystems like mangroves ecosystems (Field, 1995; Lovelock et al., 2017) that are situated in low elevated lands interfacing between sea and land on Earth's surface. It can threaten long-term sustainability of coastal communities and their unique ecosystems such as coral reefs, salt marshes and mangroves (Nicholls and Cazenave, 2010; Woodruff et al., 2013; Lovelock et al., 2015). However, mangroves forest have the ability to keep pace with sea level rise and to avoid inundation through vertical accretion of sediments (Kirwan and Megonigal, 2013).

SLR is already occurring as measured in the range of 12–22 cm during the twentieth century. The model has predicted mean sea level rise in the range from 0.09 to 0.88 m during the twenty-first century (Houghton et al., 2001) and another projection has indicated the range from 1980 to 1999 to the end of 2099 is 0.18–0.59 m (Solomon et al., 2007).

6.5.1   MEASUREMENT OF SEA LEVEL RISE

Sea level rise refers to an increase in the volume of water in the world's oceans, resulting in an increase in global mean sea level (GMSL). The average level of the sea is considered as GMSL, and it is usually measured based on hourly values taken over a period of at least a year. For geodetic purposes, the mean level may be taken over several years. The instrumental record of sea level change is mainly comprised of tide gauge measurements over the past two to three centuries and, since the early 1990s, of satellite-based radar altimeter measurements (Church and White, 2006; Church et al., 2013).

To measure sea-level change, tide gauges are used in different locations around the world in past. But, distribution of tide gauges is found uneven around the globe; this biases the data and does not give an accurate scenario of the global pattern of sea-level change (Cabanes et. al., 2001). As measured, the rate of change of relative sea-level at a tide gauge may vary greatly from the relative sea-level rate of change. It occurs in coastal wetlands due to changing elevation of the wetland sediment surface (Gilman et al., 2008). The deviation from the actual change may be due to differences in various factors such as local tectonic processes, coastal subsidence, sediment budgets, and meteorological and oceanographic events between the section of coastline in which the coastal wetland is located and a tide gauge (Gilman et al., 2008). In global sea-level change, uncertainties also prevail to the extent by which the regions may have experienced different rates and magnitudes of sea level rise. Tectonic movements which can cause land subsidence or uplift also affect regional sea-level rise (Elizabeth and Salm, 2006). However, global mean sea level rose 1.7 ± 0.5 mm yr^{-1}over the twentieth century and 1.8 ± 0.5 mm yr^{-1} from 1961 to 2003 as indicated by tide gauge measurements (IPCC, 2007). Seasonal wind directional change, for example, northwest monsoons can alter sea level by far greater amounts. For example, in the Gulf of Carpentaria, the northwest monsoon in summer raises mean sea level by about one meter, and in May to October, the southeasterly winds lower mean sea level by one meter. Nevertheless, this seasonal change in mean sea level has not affected the mangroves.

The high-precision satellite altimetry record started in 1992 and provides nearly global ($\pm 66°$) sea level measurements at 10-day intervals. Satellite altimetry since 1993 provides a more accurate measure of global sea level rise. There were measurements by three different satellites:TOPEX/Poseidon (launched 1992), Jason–1 (launched 2001) and Jason–2 (launched 2008). These data have exhibited a more-or-less steady increase in Global Mean Sea Level (GMSL) of around 3.2 ± 0.4 mm yr^{-1} over that period. The data are more than 50% higher than the average value recorded in the twentieth century (Church et al., 2013). However, it is widely accepted belief that in total, sea level rose by 6 cm during the nineteenth century and 19 cm in the twentieth century (Svetlana et al., 2008). Another estimate shows Global mean sea level (GMSL) to have increased by about 21 cm to 24 cm (8–9 inches) since 1880, with about 8 cm (3 inches) occurring since 1993 (Nerem et al., 2010; Church and White, 2011; Hay et al., 2015).

6.5.2   SEA LEVEL CHANGE AND SHIFTING OF MANGROVES

Mangroves grow well in a tidal environment where sea level is predicted to change. Mangroves need to adapt in changing sea-level for survival over long timescale. This will be a common adaptation strategy to all mangroves species. The ability of mangroves to adapt efficiently with changing sea-level, as recorded, depends on accretion rate relative to rate of sea-level change (Alongi, 2008).

The previous sea-level changes created both crises and opportunities for mangroves as indicated by the geological records. In that juncture, they survived or expanded in several refuges (Field, 1995). Mangroves may occupy laterally into areas of higher elevation. In such areas are available of suitable environmental conditions that facilitate mangroves for recruitment and establishment as befitting with relative sea-level rise (Gilman et al., 2008). These environmental conditions comprise as (i) suitable hydrology and sediment composition, (ii) competition with non-mangrove plant species, and (iii) availability of waterborne seedlings (Krauss et al., 2008).

The present climate change and sea level rise affects mangroves habitats distribution along the coastlines to a great extent. Pramanik (2015) studied recently the impact of Sea Level Rise on mangrove dynamics in the Indian Sundarban by using Geospatial Techniques. The study indicates that if the sea level rises above mangrove surface, the mangrove retreats landward; as a result the land areas decline and soil erosion increases. However, expectedly low-level mangrove habitats will be vulnerable to increasing sea level under present climate change worldwide.

6.5.3   SEDIMENT ACCRETION AND ELEVATION IN RELATION TO OCCURRENCE OF MANGROVES

Mangroves occur in locations where topography exposes them to suitable tidal or other flooding regimes (Woodroffe et al., 2016). Sedimentation within the available accommodation space can be rapid where there is abundant supply of inorganic sediment (Lovelock et al., 2010; Lovelock et al., 2015). There is a positive correlations between relative sea-level rise and mangrove sediment accretion as documented by Cahoon and Hensel (2006), which indicates that mangroves keep pace with regional relative sea-level rise. Relative sea-level rise on a particular shoreline refers to the net change caused by adjustments of the sea surface and those of the land (Woodroffe et al., 2016). Many large deltas of the world are subsiding at rates up to an order of magnitude greater than the global rate of sea-level rise (Syvitski et al., 2009, Hanebuth et al., 2013, Auerbach et al., 2015). The rate of inorganic sediment accretion may reduce exponentially with increasing sediment elevation, which occurs due to decrease of both frequency and duration of tidal inundation (Allen, 1990, 1992; French, 1991, 1993; Saad et al., 1999; Woodroffe, 2002; Cahoon and Hensel, 2006). When mangroves grow in low relief islands of carbonate settings that lack rivers, the entire vegetation is likely to be the most sensitive to sea-level rise due to their sediment-deficit environments (Thom, 1984; Ellison and Stoddart, 1991; Woodroffe, 1987, 1995, 2002). However, where large rivers supply abundant terrigenous sediments, there are allochthonous mineral sediments to accumulate rapidly in mudbanks, which provide habitats for mangroves as opportunistic colonizers (Proisy et al., 2009; Lovelock et al., 2010; Swales et al., 2015). Subsurface controls on mangrove sediment elevation can offset high or low sedimentation rates as the recent studies have shown (Cahoon et al., 2006; Cahoon and Hensel, 2006); that sedimentation rates alone give a poor indicator of fragility to sea-level rise. For instance, sediment delivery is declining in the mangroves forests of Indo-Pacific region due to anthropogenic activities (Milliman and Farnsworth, 2011; Giosan et al., 2014). This region is expected to have variable, but high, rates of future sea level rise (Chruch et al., 2013; Stammer et al., 2013) and seems to vulnerable in future. Importantly, this region holds most of the world's mangroves forests (Giri et al., 2011).

6.5.4 SEDIMENT ACCRETION IN FACILITATING MANGROVES ESTABLISHMENT

- Sediment accretion increases with increased hydroperiod, including duration, frequency and depth of inundation. The occurrence of an increased sedimentation can promote mangrove plant growth through direct effects on elevation as well as sufficient nutrient delivery. Mangroves forests may survive thousands of year in the upper tidal zone where tidal ranges are large and sediment supply are consistent (Lovelock et al., 2015). However, the capacity of soil surface to keep pace with sea-level rise is strongly dependent on the accumulation of organic matter in the low sediment supply system. And organic matter is derived from roots that decompose slowly in anaerobic soils (McKee et al., 2007).
- The increasing sediment accretion occurs through deposition of organic matter as well as increasing sediment retention with the decreasing rate of flood waters flow that would cause higher tree productivity and root accumulation (Cahoon et al., 1999; McKee et al., 2007).

- Mangroves are considered as opportunistic colonizers through their ecological responses to environmental factors (Tomlinson, 1986; Naidoo, 1985, 1990; Duke, 1992; Wakushima et al., 1994a, b; Duke et al., 1998; Cannicci et al., 2008). This indicates the importance of the geomorphic setting to determine the areas in which mangroves establish, with response to their structures and functions (Woodroffe, 2002).
- Mangrove's geomorphic setting, including sedimentation processes through sediment type and supply, hydrology, and energy regime, is an important event in responses to sea level changes.

These factors affect elevation of the mangrove sediment surface in relation to both surface and subsurface controls (Gilman et al., 2008).

- An individual mangrove experiences different trends in sediment elevation (Krauss et al., 2003; Rogers et al., 2005b; McKee et al., 2007), which indicates the differences between the change in surface elevation and adequately characterized mangroves site.
- However, there has been no significant correlation as observed between trends in mangrove sediment elevation and relative sea-level change, tidal range, or soil bulk density, even no correlations observed between geomorphic class and trends in mangrove sediment elevation (Cahoon and Hensel, 2006).

6.5.5 MANGROVES RESPONSES TO CHANGING SEA-LEVEL

When changing sea-level is the predominant factor controlling mangrove position (Figure 6.1), there are three general responses from mangroves to sea-level trends (Gilman et al., 2008):

a) Rising of site-specific relative sea-level: when sea-level rises relative to the elevation of the mangroves sediment surface, the mangroves' seaward margins migrate landward as the mangrove species maintain their preferred hydroperiod.

b) Persistence of site-specific relative sea-level: when sea-level remains stable relative to the mangrove surface, mangroves position remains persistent.

c) Subsidence of site-specific relative sea-level: when sea-level is falling relative to the mangrove surface (Krauss et al., 2014), mangroves margins migrate seaward and possibly laterally. Mangroves establish to the adjacent areas subject to suitable environmental conditions (Saintilan et al., 2014).

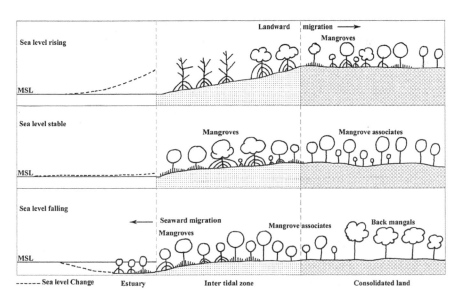

FIGURE 6.1 Predicted response of mangroves to sea level changes.

What will happen for mangroves when the seaward mangroves margin migrates landward?

• Mangroves tree may dieback due to stresses caused by a rising sea-level such as erosion resulting in weakened root structures and falling of trees, increased salinity, along with extended period of inundation, height of inundation and increasing tidal frequency (Naidoo, 1983; Ellison, 1993, 2000, 2006; Lewis, 2005).
• Mangroves migrate to new habitat (Saintilan et al., 2014) as available landward through effective seedling recruitment and vegetative reproduction, with possibility that new habitat may have different environmental settings (Alongi, 2002; Cahoon and Guntenspergen, 2010) such as erosion, inundation, and concomitant change in salinity (Semeniuk, 1994).

- Mangroves species occupy newly available habitat based on individual ability at a rate that keeps pace with the rate of relative sea-level rise (Field, 1995; Duke et al., 1998; Lovelock and Ellison, 2007), slope of adjacent land and presence of obstacles to landward migration of the landward mangrove boundary (e.g., seawalls, roads). In effect, some mangroves may gradually be reduced in area, may revert to a narrow fringe, with survival of individual trees or experience local extinction (Gilman et al., 2008).

6.5.6   MANGROVES RESISTANCE AND RESILIENCE TO RELATIVE SEA-LEVEL RISE

Scientists consider that mangroves become resistant and resilient to relative sea-level rise over a long period of geological scales. Resistance is used here to refer to a mangrove's ability to keep pace with rising sea-level without alteration to its functions, processes and structure (Odum, 1989; Bennett et al., 2005). Resilience refers to the capacity of a mangrove to naturally migrate landward in response to rising sea-level, such that the mangrove ecosystem absorbs and reorganizes from the effects of the stress to maintain its functions, processes and structure (Carpenter et al., 2001; Nystrom and Folke, 2001).

i)   Mangrove species have different ability based on time requirement to occupy new habitat that is available with relative sea-level rise. The species that occupy more quickly in habitats may outcompete slower colonizers through competitive succession and become more dominant over others (Lovelock and Ellison, 2007).

ii) Mangrove species composition is highly related to mangrove responses to sea level change. Different mangrove vegetation zones experiences different rates of change in sediment elevation (Krauss et al., 2003; Rogers et al., 2005b; McKee et al., 2007); some zones become more resistant and resilient to rising sea-level.

iii) The physiographic settings affect mangrove resistance, which includes the slope of land the mangrove currently occupies, and presence of obstacles to landward migration (Gilman et al., 2007a).

iv) Cumulative effects of all stressors influence mangrove resistance and resilience.

6.6 MANGROVES TO COPE WITH CLIMATE CHANGE

There has been little attention paid to the adaptive capabilities of mangroves to cope with disturbances (Smith, 1992; Ellison and Farnsworth, 2000). Alongi (2008) suggests that mangroves possess a variety of key attributes which contribute to their resilience to cope with climate changes or other disasters such as tsunami. These characteristics include the following:

- Nutrients are available in a large below-ground reservoir, which can replenish nutrient losses.
- Nutrient flux and microbial decomposition occurs at rapid rates, which facilitate rapid biotic turnover.
- High rates of water-use and nutrient-use efficiency through complex and highly efficient biotic controls facilitates internal reuse of resources to enhance recovery.
- Despite different species composition, self-design and simple architecture leads to rapid reconstruction and rehabilitation post- disturbances.
- Redundancy of keystone species, or species legacies leads to restoration and recovery of key forest functions and structure.
- Positive and negative feedback pathways may render malleability to help dampen oscillations recovery that reaches to a more stable and persistent state.

Smith (1992) suggested that factors like temperature, salinity (as affected by rainfall and extent of freshwater runoff), frequency and duration of tidal inundation, soil texture and degree of soil anoxia, pH, predation, and competition are also determinants of how mangroves cope with zonation. Evidently, some of these factors function synergistically and antagonistically to control mangrove distribution over different spatial and temporal scales (Alongi, 2008). Smith (1992) advocates further that mangroves have more pioneer-phase than mature-phase features: continuous production of propagules with numerous numbers, long propagule dormancy with viability, abiotic means of propagules dispersal, wide dispensability, light-demanding seedlings, dependence on seed reserves, early reproductive age, uniform crown shape, competition for light, non-specific pollinators, prolonged flowering period, tendency towards inbreeding, poor species richness, no or little canopy stratification, even-sized tree distribution, few climbers, and few epiphytes.

It is more logical than a simple argument that mangroves are likely to sustain whatever harsh environment may be due to climate changes. Mangroves already exist and adapt in harsh ecological state and survived too since paleontological ages. However, mangroves species composition, structure and function may be greatly variable in species or in communities' level and in micro or macro-environmental scale through development of species-specific adaptive strategies.

6.7   DYNAMIC CHANGE IN MANGROVES OVER GEOLOGIC TIMESCALES

Strong evidence in support of dynamic change in mangroves over geological timescales is substantiated with the existence of fossils and mangrove peat deposits in various parts of the world (Ellison and Stoddart, 1991; Plaziat, 1995; Kim et al., 2005; Alongi, 2015). Mangroves appeared and changed its position befitting with changing climate at considerable speed (Saenger 1998; Alongi, 2008), with tracing back ancestral mangroves over 65 million years (Duke, 1992). Mangroves appear to have evolved during the Cretaceous period around the fringes of Australia and New Guinea (Pernetta, 1993). It is evident that the present position of the world's mangrove forests is a legacy of the Holocene (Woodroffe, 1992, 2002; Lessa and Masselink, 2006). Alongi (2008) opined that mangrove forests are highly dynamic because of their location. Mangroves have experienced almost continual disturbance due to fluctuations in sea-level over the last few thousand years (Woodroffe, 1990, 1992; Yulianto et al., 2005). Changes in sea-level were abrupt as the fact is supported by the analyzes of sediment cores from these ancient river beds (Hull, 2005). Also, the analyzes of stratiographic sequences in mangrove peat deposits indicate that mangroves gradually moved inland while the fringing seaward stands died back as sea-level rose during the Holocene (Woodroffe, 1990; Plaziat, 1995; Kim et al., 2005). Mangroves are often pioneers as occupying newly formed mud flats and slowly move in intertidal position of what exist now in the face of environmental change (Alongi, 2008).

Mangrove responses to relative sea-level rise require few analyzes for its prediction. Gilmen et al. (2008) suggests that numerous factors other than change in relative sea-level can affect mangrove margin position, as well as structure and health. When sea-level rises keeping pace with relative to the elevation of the mangroves sediment surface, it is likely to be the predominant factor to control mangrove position. In such condition, mangroves responses will generally follow the trends, which happened in paleoenvironmental reconstructions of mangroves to past sea-level fluctuations over decades (Woodroffe et al., 1985; Ellison and Stoddart, 1991; Woodroffe, 1995; Shaw and Ceman, 1999; Ellison, 1993, 2000; Berdin et al., 2003; Dahdouh-Guebas and Koedam, 2008; Ellison, 2008; Gilman et al., 2008).

Mangroves colonize intertidal areas where dormant environmental conditions facilitate sediments to accumulate (Woodroffe, 2002; Perry, 2007). Mangroves efficiently trap fine sediment particles by slowing water movement to speed conducive for settlement of fine clay and silt particles (Wolanski, 1995). Tree trunks, roots, pneumatophores, and animals colonizing above-ground tree parts add to the friction created by the soil surface, forest floor slope, and the numerous burrows, cracks, and fissures that mark the forest floor. Furthermore, fine roots bind sediment as bacteria and other organisms living in or on the soil does the same through secretion of mucus (Alongi, 2005).

6.8 MANAGING MANGROVES

As already discussed, readers may have some idea about what climate change is? Which components of climate are responsible to climate change? What degree of intensity of climate components will affect mangroves? How will climate components disturb mangroves? How will mangroves respond to climate change?

Nevertheless, in view of present climate change, managing mangroves is a challenge. Here is a precise table presented to identify what extremes prevail in the environment due to climate change (Table 6.1), and which mangroves are to be more resilient to sustain in climate change as analyzed with selective attributes mangroves possess (Table 6.2).

6.8.1 GENERAL INFORMATION AND FEW STRATEGIC POINTS

Now, what should be the roles managers have to play to protect mangroves in changing climate? It is clearly logical argument based on past evidence that some mangroves will continue to exist by adapting in new habitats

TABLE 6.1 Climate Components, Their Expected Changes, and Mangroves Responses to Extremes in Climate Change

Climate change component	Expected changes in habitat	Mangroves response
Sea level rise	Coastal surface: • Land loss Hydrology: • Increasing tidal inundation, frequency and depth • Increasing/decreasing salinity subject to freshwater connectivity	• Mangroves may have landward migration to colonize new habitats through seedlings establishment. • Different composition of mangroves may establish a new association through competitive succession. • Mangroves which establish new habitats will become more resilient and resistant against extremes in climate change. • Secondary productivity may increase caused by nutrients availability due to land erosion and accretion.
Temperature rise	• Expansion of latitudinal range • Increasing aridity • Variable temperature in regional scale • Increasing sediment temperature • Change in atmospheric water vapor concentration	Biotic composition • Species composition may change, with increasing species diversity. Growth and physiology • Mangroves regeneration occurs caused by nutrient recycling due to bacterial growth rates in increased temperature. • Mangrove productivity will increase through normal photosynthesis and that will enhance litter production. • Productivity may increase where temperature does not exceed an upper threshold. • Length of time between flowering and the fall of mature propagules may increase that will ensure propagules supply.

TABLE 6.1 *(Continued)*

Climate change component	Expected changes in habitat	Mangroves response
Atmospheric CO_2 rise	• Increasing salinity • Increasing aridity • Change in water vapor concentration and pattern	• Increase in photosynthesis and early mangroves growth, with substantial change of vegetation when there is low salinity caused by freshwater supply. • Mangroves growth suffers when enhanced CO_2 is combined with reduced rainfall that cause high salinity and evaporative demand to restrict carbon gain at the leaves. • Flowering period may take place in advance. • In arid regions, mangroves may benefit due to the increase in water use efficiency as reduced water loss in transpiration will enhance CO_2 uptake.
Changes in precipitation	• Changes in soil salinity • Changes in soil water content • Changes in rate of evaporation • Changes in supply of fluvial sediment and nutrients	• Decreased precipitation may result in decreasing net primary productivity, growth and seedling survival. • Decreased precipitation may lead to competition among mangrove species, alteration in species composition, poor mangroves diversity. • Decreased precipitation may encourage salt-tolerant mangroves occupying land zones. • Increasing precipitation may favor higher mangrove diversity and productivity. • Increasing precipitation will increase mangroves area and diversity of mangrove zones, with the colonization of previously unvegetated areas of landward zones.

TABLE 6.1 *(Continued)*

Climate change component	Expected changes in habitat	Mangroves response
Changes in storm pattern	• High storms may alter sediment elevation through soil erosion, soil deposition, soil compression, soil salinity, flood inundation, leading to peat collapse	• Average height of mangroves will decrease, with different combination of species diversity.
		• Ecosystem setting may alter, with changes in faunal diversity.
		• Mangroves physiology and normal function is likely to be affected.
Changes in ocean circulation patterns	• Increasing ocean heat and salinity lead to changes in ocean current circulation	• Surface ocean water circulation patterns may affect mangrove propagules dispersal and thereby affecting mangroves community structure.

(Source: Naidoo, 1983, 1985, 1990; Burchett et al. 1984; Clough 1984; Saenger and Moverly 1985; Tomlinson, 1986; Hutchings and Saenger 1987; Allen, 1990, 1992; French, 1991, 1993; Duke, 1992; Pernetta 1993; Snedaker, 1993, 1995; Ellison, 1993, 2000, 2006; Smith et al., 1994; UNEP 1994; Wakushima et al., 1994a, b; Field 1995; Duke et al., 1998; Saad et al., 1999; Woodroffe and Grime, 1999; Baldwin et al., 2001; Houghton et al. 2001; Saintilan and Wilton, 2001; Sherman et al., 2001; Church et al. 2001, 2004; Woodroffe, 2002; Krauss et al., 2003; Ning et al. 2003; Cahoon et al., 2003, 2006; Harty 2004; Rogers et al., 2005a; Danielsen et al., 2005; Ellison 2005; Kathiresan and Rajendran, 2005; Lewis, 2005; Rogers et al., 2005a, 2005b; Whelan et al., 2005; Dahdouh-Guebas et al., 2005a, b, 2006; Cahoon and Hensel, 2006; Cahoon et al., 2006; Gilman et al. 2006; McLeod and Salm, 2006; Piou et al., 2006; Lovelock and Ellison, 2007; McKee et al., 2007; Solomon et al., 2007; Alongi, 2008; Cannicci et al., 2008; Gilman et al. 2008; Krauss et al., 2008).

through developing resilience or by migration to colonize new areas and perhaps even thrive with the predicted changes in climate. But it is also obvious that some won't exist as happened in the past. Nevertheless, managers will conceive before what will happen for mangroves due to predicted climate change. They will need to plan strategically to protect all mangroves species in respective environmental setting of regional scale in such critical juncture, particularly sea level rise. We like to mention some steps for managing mangroves in changes in climate.

i)   To identify representative species, along with respective habitats.
ii) To observe the key attributes of representative species and identify whether they are able to sustain in harsh environmental habitat due to predicted climate change.
iii) To increase seedlings of particular species through silviculture practices.
iv) To observe what conditions the seedlings require for its survival.
v)   To plan how to protect its habitats.
vi) To know what habitats it prefers to grow in, for example riverine, basin or fringe habitats.
vii) To observe what salinity range it is able to grow in, apart from considering sea level, tidal fluctuation and tidal range.
viii) To identify in which zonation it grows and survives normally.
ix) To study with which species association, it grows naturally in mutual competition.
x)   To segregate mangroves based on their adaptability in befitting with respective habitats.
xi) To maintain biodiversity so as to ensure that a wide range of biota may survive and become associated to be effective for ecosystems function.\
xii) To compensate immediately, if any species disappear in its normal habitats and to determine the reasons for its disappearance.

6.8.2 STRATEGIC PLANS FOR BUILDING MANGROVES RESILIENCE

Some strategic plans are suggested for mangroves to have resiliency to climate change. They are listed in the following subsections.

6.8.2.1 HABITATS

- Identify habitats which are vulnerable to climate change and protect them from future danger. In such case, manager needs to identify areas which are naturally positioned to withstand against climate change. Apply such strategy, establish and spread it to address the uncertainties of climate change.
- Apply alternative strategy to protect mangroves to escape the danger from sea-level change, when protected area managers are unable to control large-scale such threats. For example, managers can identify alternative places adjacent to the habitats to re-establish it for mangroves plantation if natural recruitment does not occur.
- Facilitate mangrove migration as much as possible in response to sea-level rise through establishment of greenbelts and buffer zones and protect mangroves by reducing impacts arising out of adjacent land-use practices.
- Make degraded areas restored which have proven resistance or resilience to climate change. Have understanding to establish connectivity between mangroves and sources of freshwater and sediment, and between mangroves and their adjacent habitats like sea grasses and coral reefs.

6.8.2.2 MANGROVES

- Establish strategies which are to compensate for alteration in species composition and changes in environmental conditions (Table 6.2; Figure 6.2).
  - i)   Select species tolerant to high salinity.
  - ii) Select species resistant to strong wind, sea surges, hurricanes and sea wave.
  - iii) Select species resilient to soil erosion and expand immediately through seedlings' recruitment (Figures 6.3 and 6.4 show mangroves resilience in environmental adversities).
  - iv) Select species persistent to deep submergence with long duration.
  - v)   Select species more buoyant to escape sea wave and float far and long.
  - vi) Select species having long dormant period, with effective viability.

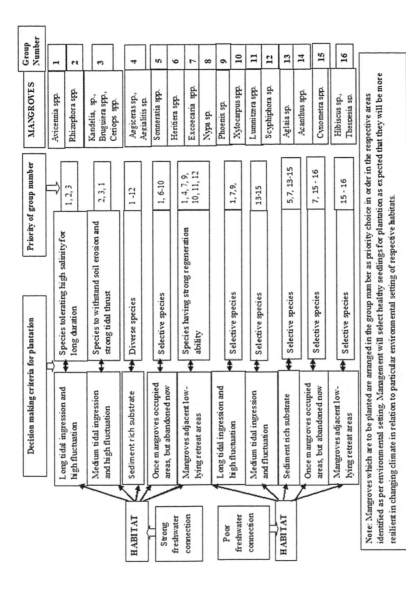

FIGURE 6.2 Management options for building mangroves' resilience to climate change.

FIGURE 6.3 (a) A *Rhizophora* spreads stilt roots to withstand against adversity; (b) aerial roots (pneumatophore) show unusual growth in anaerobic substrate; (c) a number of *Rhizophora* plants develop dense stilt roots for survival against tidal thrust; (d) *Avicennia* develops pneumatophores in anaerobic substrate; (e) *Excoecaria agallocha* spreads serpentine roots in eroded soil bed; (f) *Bruguiera gymnorhiza* develops knee roots for breathing in inundated zone.

6.8.2.3 MANAGEMENT

- Manage human stresses on mangroves through awareness and education.
- Monitor the response of mangroves to climate change through establishment of baseline data.
- Compensate for changes in species ranges and environmental conditions by implementation of adaptive strategies.
- Minimize mangrove resources dependency among locals through development of alternative livelihoods as a means for mangrove-dependent communities.
- Generate the necessary finances and supports to develop partnerships among a variety of stakeholders to respond the impacts of climate change.

6.8.3 EFFECTIVE MANAGEMENT

McLeod and Salm (2006) suggest some strategic steps of effective management for mangroves. We modify them and like to reorient all these into few strategic groups for effective management as given in the following subsections.

6.8.3.1 CONSIDERATION OF ENVIRONMENTAL SETTING AND RELATED CONDITIONS

The factors determining environmental setting in response to threats from climate change include forest width, forest floor slope, forest location (open coast vs. lagoon), type of adjacent lowland vegetation and cover, soil texture, presence of foreshore habitats (sea grass meadows, coral reefs, dunes), size and speed of tsunami, distance from tectonic event, and angle of tsunami incursion relative to the coastline, tree density, tree diameter, proportion of above-ground biomass, roots, and tree height.

Managers need to minimize land-based threats to mangroves by developing land-use practices to check nutrient and sediment run-off, prohibit illegal wood cutting, control the use of persistent pesticides, and regulate strict filtration of effluent to make water quality better.

6.8.3.2   ESTABLISHMENT OF GREENBELTS AND BUFFER ZONES

Mangroves greenbelts may provide effective protection of shoreline from erosion and need to be established along erosion-prone estuarine areas, coastlines, river banks and in areas which are vulnerable to sea surges, sea waves, tidal surges, tsunami, cyclones, and geomorphic erosion (Macintosh and Ashton, 2004). Greenbelts need to be developed a minimum of 100 m, but preferably up to 500 m or 1 km (McLeod and Salm, 2006).

Buffer zones which border the seaward and landward margins of protected mangrove areas are to be established to facilitate a transition between human settlements with intensively used lands and waters and the protected area. For a variety of mangroves, the landward zones are vital that experience sea-level rise to facilitate landward expansion. The land-use practices around buffer-zones need to be "biodiversity-friendly" as to keep it distant from pollution or to be protected from adjacent contaminated areas like pesticide used farming (Barber et al., 2004).

6.8.3.3   RESTORATION OF DEGRADED AREAS

Mangrove areas which are potential habitats for mangroves but degraded recently need to be restored. Determining the effective criteria for restoration of degraded lands, some features are mentioned as;

- Having adjacent active drainage systems including permanent rivers, channels and creeks that supply freshwater and sediment.
- Lands with sediment rich condition to facilitate sediment deposition, redistribution and accretion.
- Progradation of coast and delta.
- Provision of effective hydrological restoration with cost-effective approach includes two main types of hydrological restoration: (i) restoring tidal hydrology through excavation or backfilling, and (ii) rechanneling blocked areas to normal tidal influences (Lewis and Streever, 2000).
- Natural protected zones such as bays, barrier islands, beaches, sandbars, and reefs that decrease wave erosion and storm surge.

6.8.3.4 ESTABLISHMENT OF MANGROVES THROUGH PLANTATION

An aggressive mangroves plantation program considering the degree of resilience of respective species (Table 6.2; Figure 6.2) should be undertaken to develop resilience to climate change, apart from protection of existing mangroves forest from anthropogenic threats. Healthy mangroves are resilient well to withstand climate changes (McLeod and Salm, 2006). Major areas of mangroves forest are located within 25 km of urban areas which are usually inhabited by 100,000 or more people (Millenium Ecosystem Assessment, 2005). More is the density of mangroves, much will be their resilient to climate changes. It is seen that substantial mangroves areas are located near to urban areas and are likely to receive threats such as pollution, diking, channelization, illegal felling, conversion to rehabilitation, aquaculture, agriculture, and other forms of coastal development.

FIGURE 6.4 *Rhizophora* plant stands tall by dint of stilt roots against tidal thrust.

TABLE 6.2　Different Mangroves Species Assessed With Degree of Resiliency Based on Their Respective Attributes and Adaptive Ability

Sl. no.	Scientific names of mangroves	Organ	Attributes	Adaptive ability	Degree of resiliency
1	*Acanthus ebracteatus* Vahl	Fruit	Crypto-vivipary	Fruit coat buoyant and float long duration	Low resilient
2	*Acanthus ilicifolius* L.	Root	Stilt roots	Withstand against soil erosion	High resilient
		Leaf	Salt gland	Remove excess salt while growing in high salinity	
		Fruit	Crypto-vivipary	Fruit coat buoyant and float long duration	
3	*Acrostichum aureum* L.	Spore	Enormous numbers	High dormancy	Low resilient
4	*Aglaia cucullata* (Roxb.) Pellegr.	Fruit	Buoyant	Float far with long duration	Low resilient
5	*Aegialitis rotundifolia* Roxb.	Root	Aerial roots	Withstand against soil erosion	High resilient
		Leaf	Salt gland	Remove excess salt while growing in high salinity	
		Fruit	Crypto-vivipary	Fruit coat buoyant and float long duration	
6	*Aegiceras corniculatum* (L.) Blanco	Leaf	Salt gland	Remove excess salt while growing in high salinity	High resilient
		Fruit	Crypto-vivipary	Fruit coat buoyant and float long duration	
7	*Avicennia alba* Blume	Root	Pneumatophore	Continue physiology in highly anaerobic substrate	High resilient
		Leaf	Salt gland	Remove excess salt while growing in high salinity	
		Fruit	Crypto-vivipary	Fruit coat buoyant and float long duration	
8	*Avicennia marina* (Forssk.) Vierh.	Root	Pneumatophore	Continue physiology in highly anaerobic substrate	High resilient
		Leaf	Salt gland	Remove excess salt while growing in high salinity	
		Fruit	Crypto-vivipary	Fruit coat buoyant and float long duration	
9	*Avicennia officinalis* L.	Root	Pneumatophore	Continue physiology in highly anaerobic substrate	High resilient
		Leaf	Salt gland	Remove excess salt while growing in high salinity	

TABLE 6.2 (Continued)

Sl. no.	Scientific names of mangroves	Organ	Attributes	Adaptive ability	Degree of resiliency
10	*Bruguiera cylindrica* (L.) Blume	Fruit	Crypto-vivipary	Fruit coat buoyant and float long duration	High resilient
		Leaf	Thick hypodermal tissue	Store excess salt	
		Fruit	Vivipary	Hypocotyl having buoyant ability to float far and long duration with viability	
11	*Bruguiera gymnorhiza* (L.) Savigny	Root	Knee and buttress	Withstand against soil erosion	High resilient
		Leaf	Thick hypodermal tissue	Store excess salt	
		Fruit	Vivipary	Hypocotyl having buoyant ability to float far and long duration with viability	
12	*Bruguiera parviflora* (Roxb.) Wight &Arn. Ex Griff.	Root	Buttress	Withstand against soil erosion	High resilient
		Leaf	Thick hypodermal tissue	Store excess salt	
		Fruit	Vivipary	Hypocotyl having buoyant ability to float far and long duration with viability	
13	*Bruguiera sexangula* (Lour.) Poir.	Root	Buttress	Withstand against soil erosion	High resilient
		Leaf	Thick hypodermal tissue	Store excess salt	
		Fruit	Vivipary	Hypocotyl having buoyant ability to float far and long duration with viability	

TABLE 6.2 *(Continued)*

Sl. no.	Scientific names of mangroves	Organ	Attributes	Adaptive ability	Degree of resiliency
14	*Cerbera odollam* L.	Fruit		Buoyant ability	Low resilient
15	*Ceriops decandra* (Griff.) Ding Hou	Root	Broom like stilt roots	Withstand against soil erosion	High resilient
		Leaf	Thick hypodermal tissue	Store excess salt	
		Fruit	Vivipary	Hypocotyl having buoyant ability to float far and long duration with viability	
16	*Ceriops tagal* (Perr.) C.B. Rob.	Root	Knee and buttress	Withstand against soil erosion	High resilient
		Leaf	Thick hypodermal tissue	Sore excess salt	
		Fruit	Vivipary	Hypocotyl having buoyant ability to float far and long duration with viability	
17	*Cynometra ramiflora* L.	Fruit		Buoyant with long dormancy	Low resilient
18	*Excoecaria agallocha* L.	Root	Much spreading	Soil binding ability	High resilient
		Fruit	Small, numerous and buoyant	Floating ability with long dormancy	
19	*Heritiera fomes* Buch.-Ham.	Root	Pneumatophore and ribbon like Buttress	Continue physiology in highly anaerobic substrate and withstand against soil erosion	Low resilient
		Fruit	Buoyant	Float, but short dormancy and cannot tolerate high salinity	
20	*Heritiera littoralis* Aiton	Root	Flat and long buttress	Continue physiology in highly anaerobic substrate and withstand against soil erosion	Medium resilient

TABLE 6.2 (Continued)

Sl. no.	Scientific names of mangroves	Organ	Attributes	Adaptive ability	Degree of resiliency
		Fruit	Buoyant	Float, but short dormancy and can tolerate high salinity	
21	*Hibiscus tiliaceus* L.	Fruit	Buoyant	Buoyant and floating	Medium resilient
22	*Kandelia candel* (L.) Druce	Root	Broom like stilt roots	Withstand against soil erosion	High resilient
		Leaf	Thick hypodermal tissue	Store excess salt	
		Fruit	Vivipary	Hypocotyl having buoyant ability to float far and long duration with viability	
23	*Lumnitzera littorea* (Jack) Voigt	Root	Spreading	Soil binding ability	Medium resilient
		Leaf	Succulent	Accumulate excess salt so as to tolerate high salinity	
24	*Lumnitzera racemosa* Willd.	Root	Spreading	Soil binding ability	Medium resilient
		Leaf	Succulent	Accumulate excess salt so as to tolerate high salinity	
25	*Nypa fruticans* Wurmb	Fruit	Crypto-vivipary	Buoyant with long dormancy and viability, but grow in low salinity	Low resilient
26	*Phoenix paludosa* Roxb.	Root	Pneumatothod	Develop in highly anaerobic substrate	High resilient
		Fruit	Small numerous	Buoyant with long duration	

TABLE 6.2 *(Continued)*

Sl. no.	Scientific names of mangroves	Organ	Attributes	Adaptive ability	Degree of resiliency
27	*Rhizophora apiculata* Blume	Root	Stilt roots	Withstand against soil erosion	High resilient
		Leaf	Thick hypodermal tissue	Store excess salt	
		Fruit	Vivipary	Hypocotyl having buoyant ability to float far and long duration with viability	
28	*Rhizophora mucronata* Lam.	Root	Stilt roots	Withstand against soil erosion	High resilient
		Leaf	Thick hypodermal tissue	Store excess salt	
		Fruit	Vivipary	Hypocotyl having buoyant ability to float far and long duration with viability	
29	*Rhizophora stylosa* Griff.	Root	Stilt roots	Withstand against soil erosion	High resilient
		Leaf	Thick hypodermal tissue	Store excess salt	
		Fruit	Vivipary	Hypocotyl having buoyant ability to float far and long duration with viability	
30	*Scyphiphora hydrophylacea* C. F. Gaertn.	Root	Spreading	Withstand against soil erosion	Medium resilient
		Leaf	Succulent	Accumulate excess salt	
31	*Sonneratia alba* Sm.	Root	Pnematophore	Continue physiology in anaerobic substrate	Medium resilient
		Leaf	Succulent	Accumulate excess salt	

TABLE 6.2 *(Continued)*

Sl. no.	Scientific names of mangroves	Organ	Attributes	Adaptive ability	Degree of resiliency
32	*Sonneratia apetala* Buch.-Ham.	Root	Pnematophore	Continue physiology in anaerobic substrate	Medium resilient
		Leaf	Succulent	Accumulate excess salt	
33	*Sonneratia caseolaris* (L.) Engl.	Root	Pnematophore	Continue physiology in anaerobic substrate	Low resilient
		Leaf	Succulent	Accumulate excess salt	
34	*Sonneratia griffithii* Kurz	Root	Pnematophore	Continue physiology in high anaerobic substrate	Medium resilient
		Leaf	Succulent	Accumulate excess salt	
35	*Thespesia populnea* (L.) Sol. ex Corrêa	Fruit	Small	Buoyant for long duration	Low resilient
36	*Xylocarpus granatum* J. Koenig	Root	Pneumatophore	Continue physiology in anaerobic substrate and withstand against soil erosion	Medium resilient
		Fruit	Buoyant	Float far with viability	
37	*Xylocarpus moluccensis* (Lam.) M. Roem.	Root	Pneumatophore	Continue physiology in anaerobic substrate and withstand against soil erosion	Medium resilient
		Fruit	Buoyant	Float far with viability	

6.8.3.5 MAINTENANCE OF CONNECTIVITY BETWEEN MANGROVES AND ASSOCIATED SYSTEMS

Maintenance of connectivity between mangroves and upland water catchments is essentially required to establish an adequate supply of sediment and freshwater. Healthy mangroves seedlings are to be selected on the assumption that if the existing areas become vulnerable to climate change, they may escape the threats and would be able to colonize new areas.

Coastal areas such as reef and fisheries which have close proximity to mangroves are to be established as pro-mangroves areas based on their connectivity (Mumby et al., 2004). Areas which seem to benefit mangroves and adjoining ecosystems by actively filtering sediments and drain out pollutants or developing nursery habitats should be protected on a priority basis. Mangroves have capability of stabilizing sediments and trap heavy metals and nutrient-rich run-off, thus improving the water quality for sea grasses, corals, and fish communities.

6.8.3.6 DEVELOPMENT OF SUSTAINABLE AND ALTERNATIVE LIVELIHOODS FOR MANGROVE DEPENDENT COASTAL PEOPLE

Unless alternative sources of livelihood are arranged dependency of coastal people on forest resources will continue (Figure 6.5 shows engagement of local mass in fishing; Figure 6.6 shows a view of variety of livelihoods in the Sundarban forest). Government sectors, particularly forest department, fishery departments in some cases, and several NGOs involved in the protection of forest need to be thinking seriously how to frame appropriate policy for uplifting socio-economic standard of forest dwellers so that their dependency on forest resources may be reduced (Das and Mandal, 2016). Development of alternative livelihoods which are less destructive than over-harvesting of mangroves or conversion to fish or shrimp ponds should be selected and established to mitigate mangrove deforestation. Plantation of mangroves in adjacent areas as restoration program may develop sustainable livelihoods for local inhabitants, thus decreasing the pressure on adjacent mangroves areas. Planting various mangrove species to create forest areas is directly correlated to restoration of diverse mangroves (Ellison, 2000), which may benefit locals for

FIGURE 6.5 A diverse age group of local mass are engaged in fishing: A way of livelihood in the Sundarban.

FIGURE 6.6 View of livelihoods: (a) fuel collection from jungle; (b) fishing in estuary; (c) fuel collection during high tide; (d) country boat used for fishing; (e) fishing by folk woman; (f) one local fisherman waits for sale of crabs caught from mangrove estuary.

livelihood. Mangroves systems of various species combination appear to be better ecological resilient (Blasco et al., 1996).

Alternative livelihoods may have some revenue generated activities such as coconut shells used for charcoal production instead of burning of mangroves, honey harvest from setting up of artificial honeycomb instead of damaging mangrove forest along with pollinators in traditional way. All these activities ensure to promote agroforestry and conservation of existing mangrove forests (Nathanael, 1964; Bandaranayake, 1998; Mandal et al., 2010; Das and Mandal, 2016).

6.8.3.7   PROVISION OF BASELINE DATA AND MONITORING PLAN

Establishment of baseline data for storing information about mangroves and projected threats is timely, urgent and essential. Data that will maintain information should include a range of variables including: tree stand structure, tree abundance, species richness, and species diversity; invertebrate abundance, richness, and diversity; primary production (biomass and litter), nutrients export; hydrologic patterns (Ellison, 2000; 2008); and rates of sedimentation and relative sea-level rise (McLeod and Salm, 2006), along with their ecological interaction.

To understand the degree of resistance of mangrove protected areas in present and future threat, mangrove ecosystems need to be monitored continuously to measure the effects of global warming such as sea level rise and human activities like over-exploitation. Changes in seashore chemistry (CO_2 levels and salinity), hydrography (sea level, currents, vertical mixing, storms and waves), tidal fluctuation, range, duration, salinity, and temperature are also to be monitored covering long timescales to determine possible climate trends and its changes (McLeod and Salm, 2006). All these available data may predict how the resilience of mangroves protected areas are befitted to withstand present and future threats. Based on the analyzes about the status of mangroves areas in respect of climate change, flexible strategies should be developed in a given boundary and monitored continuously to set up adaptive management (McLeod and Salm, 2006).

6.8.3.8   ESTABLISHMENT COLLABORATION PARTNERSHIPS AT LOCAL, REGIONAL, AND GLOBAL SCALES

A potential collaboration need to strengthen between aid agencies and conservation groups. For this, effective and strong leadership is essential to develop assured support at local, regional, and global levels. Global climate change which appears to be the greatest challenge to mangroves requires innovative collaboration for solution (McLeod and Salm, 2006). Building network among global, regional, and local partnerships covering industries such as agriculture, tourism, water resource management is required for infrastructure development for conservation. Combination of all these stakeholders with concerted effort can help reduce financial burdens and thus may respond to climate change as a large-scale threat (Shea et al., 2001).

6.8.3.9   DISSEMINATION OF MANGROVES VALUES AMONG LOCALS THROUGH EDUCATION

Mangroves support coastal people in various ways. However, most of the beneficiaries are not aware of these supports or are compelled to ignore it because of lack of knowledge. So, it is difficult to apprehend the needs of locals and then convince them for sustained use of the mangroves. Initially, values of mangroves ecosystem should be assessed, determined and then communicated at the local and national level to strengthen support for mangrove conservation (McLeod and Salm, 2006). Many tools and methods are available to encourage managers to develop resilience into their strategies formulation for mangrove conservation. In such case one can track the geological evidence of mangrove forests response to climate change (McLeod and Salm, 2006).

6.9   METHODS AND TOOLS USED FOR MONITORING AND BUILDING MANGROVES RESILIENCE

There are several tools with techniques used to monitor and build mangroves resilience, which are usually in two types based on their efficiency of prediction in relation to short-term and long-term assessment relative to climate change as Low Tech Approach and High Tech Approach.

6.9.1   LOW TECH APPROACH

- Low-tech approach that can determine fragility to sea-level rise through high-resolution satellite imageries and Geographic Information Systems (GIS) needs to be used to overlay scenarios of sea-level rise with elevation and coastal development data. Then this approach is to identify vulnerable areas (Klein et al., 2001) with much clearly and accuracy.
- Low-tech approach that can determine salinity changes and other water quality in mangrove systems is a network of piezometer clusters to be installed at the site for continuous and manual measurements of water level and salinity (Drexler and Ewel, 2001).
- Low-tech approach that determines elevation changes may be used for annual measurements of the soil elevation deficit (elevation change minus sea-level rise) which will help determine mangrove ecosystems fragility to sea-level rise (Cahoon and Lynch, 1997). Artificial soil marker horizon plots are used to measure current rates of sedimentation. Also, marker horizons can determine vertical accretion which includes both sediment deposition and sediment erosion. Besides, mangrove vertical accretion and subsidence is monitored by using Surface Elevation Tables (SETs) which are also able to provide highly accurate and precise measurements (+/– 1.4 mm total error) of sediment elevation relative to sea-level change (Cahoon et al., 2002a).

6.9.2   HIGH TECH APPROACH

- Surface elevation table-marker horizon (SET-MH) is used as high-tech approaches to measure changes in the soil surface elevation overtimes and thus to predict mangrove responses to sea-level rise (Cahoon et al., 2002c; Callaway et al., 2013). This method has been widely used and recommended for monitoring surface – elevation trajectories in coastal wetlands (Webb et al., 2013). It is also capable of providing useful information of short-term wetland elevation dynamics. However, they have several limitations which are overcome by site-specific computer models (Cahoon et al., 2002c; McLeod and Salm, 2006).

- The changes in species composition of mangroves forests through the Holocene were determined by analyzes of pollen data from the core that indicated the changes in different times because of disturbances from hurricanes, storms, change in temperature, or fluctuations in sea level. Variations in the stable carbon and nitrogen isotopes were analyzed in the fossilized mangrove leaves to determine the changes in stand structure, which is related to changes in salinity, nutrient status, and sea level (McLeod and Salm, 2006).

KEYWORDS

- climate change
- managing mangroves
- sea level rise
- threats

CHAPTER 7

MANAGEMENT AND CONSERVATION OF MANGROVES

7.1 THE VALUES OF MANGROVES

Mangroves vegetation is of immense values which usually refer to a wide variety of benefits harnessed out of ecosystem services. Nevertheless, people have common queries of why mangroves need to be protected. What service do mangroves cater to coastal zone areas? Which benefits do they render to human beings? All these queries are apparently not to be the subject related to the existence of mangroves, but appear to be mere inquisitiveness to those involved in mangroves protection, at least to planners who are supposed to care for this unique coastal vegetation. To resolve the queries, Saenger (2002), in his classical monograph on mangroves, highlighted many facets of value that mangroves provide, and categorized them broadly into four groups: (i) economic, products value; (ii) usefulness, ecosystem value; (iii) intrinsic, natural value; and (iv) symbolic, cultural and mythical value.

Ecosystem services occur through the interaction between biotic and abiotic components of a given area, resulting in generating goods and services, including nutrients cycling, flow of energy and other materials, considered as production of ecosystem functions. Provision of goods and services to people has been assessed in terms of value by many workers with times in mangroves ecosystems. In fact, not all mangroves provide all the goods and services attributed to them in worldwide, but ecosystem services of mangroves are of similar pattern irrespective of forest. Use of goods and services of mangroves forest occurs in a variety of purposes, and seems to be almost regional specific (Salem and Mercer, 2012). This field guide plans to classify all the goods and services rendered by mangroves broadly into three uses such as: (i) direct use: it refers to consumptive and non-consumptive uses that involve into direct physical interaction with the mangroves and their services, (ii) indirect use: it refers to regulatory ecological function, leading to indirect benefits, and (iii) symbolic use: it refers to existence and bequest values of mangroves (Table 7.1).

The list of benefits is believed to satisfy field workers for why to conserve mangroves, and what consequences coastal people face, if mangroves lost. Despite the general benefits listed above, there are some specific accounts on usefulness of mangroves recorded. Mangroves forests are the economic foundation of many tropical coastal regions in worldwide, with the support of at least US$ 1.6 billion per year through 'ecosystem services' (Field et al., 1998; Ewel et al., 1998). It is estimated that almost 80% of global fish catches are directly or indirectly dependent on mangroves (Sullivan, 2005; Ellison, 2008). Mangroves can sequester up to 25.5 million tons of carbon per year (Ong, 1993), and provide more than 10% of essential organic carbon to global oceans (Dittmar, 2006), occupying only 0.12% of the world's land area (Dodd and Ong, 2008). Within 150 km of coastline, almost half (44%) of the world population live (Cohen et al., 1997) and at least 40% of animal species restricted to mangroves habitats are verge of an elevated risk of extinction due to habitats loss of mangroves (Luther and Greenburg, 2009). The loss of individual mangrove species and associated ecosystem services have a direct economic effect on human livelihoods, directly related to missing of various economic fauna and indirectly leading to poor ecosystem services.

7.2   RESTORATION

This book considers the term 'Restoration' to mean any process undertaken that aims to return a system to a pre-existing condition. In mangroves restoration, the process includes natural restoration or recovery of mangroves through strategic approach. On the other hand, the term "rehabilitation" is applied to indicate any activity (including restoration and habitat creation) that aims to convert a degraded system to a stable alternative (Lewis, 1990). Saenger (2002) pointed out that some loss, degradation and alteration of mangroves occurred in past mostly because of human intervention. Given the list of mangroves ecosystem services (Table 7.1) which are considered to play immense benefits to human beings, it is necessary to undertake

TABLE 7.1 Ecosystem Services of Mangroves

Value type	Ecological function	Goods and services	Products/ Resources	Purpose
Direct use	Habitat for plant and animal species	Collection of natural products	Firewood	i. Fuel: cooking, heating, burning bricks, Smoking fish, Smoking sheet rubber, charcoal, alcohol
			Timber	ii. Construction: scaffolds, bridged, railroad ties, mining pit props, deck piling, beam and poles for building, flooring, paneling, boat building, fence posts, water pipes, chip boards, glues
			Leaves	iii. Thatching: roof, wall and floor
			Tender branches and leaves	iv. Feed: Fodder for livestock
				v. Manure: Green manure for Agriculture
			Bark and leaves	vi. Paper: Production of paper of various kinds, rayon, cigarette wrappers
			Wood and timber	vii. Households: Furniture, lumber, glue, hairdressing oil, tool handles, rice mortar, toys, matchsticks, incense, packing boxes
			Bark	viii. Textile and leather: Fibers, dye, tannin
			Leaves, bark, fruits and seeds	ix. Foods: Sugar, sweetmeats from propagules, vegetables, fruits, cooking oil
				x. Drug: Medicine
				xi. Beverages: alcohol, vinegar, tea substitute, fermented drinks, desert topping, condiments from bark,
		Commercial/recreational fishing and hunting	Fish	i. Food (Enormous fish species)

TABLE 7.1 *(Continued)*

Value type	Ecological function	Goods and services	Products/ Resources	Purpose
			Crustaceans	ii. Food (Prawn, shrimp, crab)
			Mollusks	iii. Oysters, mussels, cockles
			Bees	iv. Honey, wax
			Birds	v. Food, feather
			Reptiles	vi. Skin, food
			Mammals	vii. Food, fur
			Amphibians	viii. Food
			Insects	ix. Food, pollination
			Biological diversity and water body	x. Ecotourism: Boating, fishing, hunting,
				xi. Water transport: Transport of various goods
				xii. Recreation and esthetic beauty
		Energy resources	Biotic and abiotic components	i. Flow of energy
Indirect use	Water quality maintenance and nutrients retention	Trap sediments	Soil particles	i. Accretion of land for protecting shoreline and cope with effective sea level rise
		Process of nutrients and organic matters	Litter and soil minerals	ii. Increasing productivity of habitats
		Provision of detritus	Dead and decayed substances	iii. Addition of nutrients to enhance productivity

TABLE 7.1 *(Continued)*

Value type	Ecological function	Goods and services	Products/ Resources	Purpose
		Act as sink	Soil	iv. Nutrients and carbon
		Water cycling	Water	v. Ecosystem sustainability
	Storm buffering and sediment retention	Buffering against calamities	Vegetation	i. Calamities: Tsunami, storm, sea surges, high wind, flood, soil erosion, strong wind, high tidal thrust
	Carbon sequestration	Photosynthesis	Vegetation	ii. Carbon sequestration
	Waste disposal	Sewage treatment	Vegetation	iii. Pollution: Accumulation of pollutants
	Air quality	Evapotranspiration	Vegetation	iv. Regulation of local climate
	Pollination	Transfer of pollen	Insects	v. Provision of increase of progenies for sustainable biodiversity
	Food chain	Provision of foods	Leaves, bark, branches, fruits, seeds, shelters	vi. Breeding and nursery ground of a variety of fishery species
				vii. Provision of food and shelter to aquatic animals through food chain
				viii. Food and shelter for terrestrial animals
Symbolic use	Biodiversity	Resilience	Vegetation	i. Sustainability of coastal forest and mangroves
		Spiritual attachment	Forest and animals	ii. Involvement of communities to perform particular rituals
		Religion attachment	Forest and animals	iii. Mythological evidence related to certain communities perform worship of deities.

mangroves plantation to restore such fragile vegetation to protect them from further loss.

There is almost every species of mangroves assessed to be threatened in Red List of Threatened species (Table 2.3 and Chapter 5); 4 mangroves species out of 65 are already assessed critically endangered and endangered categories. They are likely to be extinct, if immediate remedial measures for their conservation are not undertaken. Given the vulnerability of mangroves forest in general and few mangroves set in threat in particular, the restoration of mangroves is imperative today. Saenger (2002) necessarily pointed out that practical objectives need to be clearly set up before taking any restoration effort, which must be local people friendly where this program are to be undertaken.

In this context, PRA (Participatory Rural Appraisal) may be ideal tool to implement any type of restoration program. PRA is an approach that aims to incorporate the knowledge and opinions of rural people in the planning and management of development projects and programs. This book is to guide on how to proceed towards PRA.

- Call a meeting among local people in the area in which mangroves plantation is to be undertaken.
- Brief them about the importance of mangroves with chats and illustrations, and aware them what harms happen if mangroves lost.
- Observe their feelings about mangroves, particularly restoration program.
- Form four to five groups among participants.
- Make one leader in each group.

Then separately meet each group and share their knowledge about mangroves.

Causes of mangroves disappearance

- Why was the mangroves damaged?
- How was mangrove destroyed?
- What was the look of original mangrove forest?
- How did the community treat the mangroves?

Land issues

- Who presently owns the land or has use rights to the land?
- Is the area currently used for any other purpose?

- Are the occupants willing to use the land for mangroves plantation?
- Do they guess any political interference if restoration initiates?

Environmental issues

- What are the tidal levels of this region, including height, duration and frequency?
- Where does the water come from that inundate mangrove areas?
- What are the salinity levels?
- Which are the mangroves suitable to be planted?

Livelihood issues

- What is their present livelihood?
- How do they benefit from forest?
- Which mangroves are more beneficial to them and why?
- What resources do they collect from forest?
- Do they have any idea about usefulness of mangroves?

Attitude of locals towards mangroves plantation

- How will the community treat these mangroves once they are restored?
- Which programs will be allowed/disallowed in the mangrove area?
- Who will monitor village regulations required for mangrove protection and its sustainable use?
- Is there an option for co-management with the government?
- How will people restrict outside developers/investors into the mangroves areas?
- How will people restore the mangrove areas?
- Who are the local managers interested in the mangrove restoration?

Collate all the responses together and cross-check the validity of all the information collected separately. Segregate those based on similarities and dissimilarities. Analyze which are the positive responses and which are negative ones. Find if negative responses have any impact on mangroves restoration. Try to turn negative responses to be positive ones. After doing all the pros and cons of responses, set the objectives of the restoration activities. Sanger (2002) suggested that objectives be clearly defined and prioritized as the first step in the planting process. Two distinct objectives may be addressed as;

i. **Economic benefit of the local people:** This objective needs to be formulated in such a way that will certainly fulfill the needs of local people related to timber production, fire wood for fuel, thatched materials, minor construction-timber and fodder for livestock. These issues are obviously related to social and political matters in the area where the restoration is to be undertaken, and need to be carefully managed. Once this objective is fulfilled, restoration activity is expected to accelerate with top speed.

ii. **Ecological restoration of mangroves:** It has a long impact that both directly and indirectly benefits communities as well as coastal environment. Shoreline protection, nutrient cycling, buffering natural calamities, restoration of fisheries and wildlife, all these are direct benefits harnessed out of ecological restoration, apart from livelihood support of local people dependent on forest resources.

7.3 PLANTATION APPROACH

There are many different techniques and methods used in plantation of mangroves. However, we suggest SWOT analyzes (Table 7.2) to be an ideal strategic approach before taking any plantation program. How far the proposed plantation approach is to be effective may be evaluated through SWOT (Strength, Weakness, Opportunity and Threat) analyzes. These analyzes enable the planner to understand the gap between the plan and the execution of activities.

The SWOT analysis is treated as a mirror through which plantation approach may be viewed for what item needs to have extra care and what to be strengthened; because there are reports of identifiable failures in restoration in parts of the world. This exercise helps planners understand the priorities of local people to the need of a specific mangrove or a number of mangroves species. Once their views are prioritized, their interest in plantation may rise towards active participation and continue to protect the mangroves.

TABLE 7.2   SWOT Analyses of Mangroves Plantation

Criteria	Items
Strength	i. Availability of seeds/seedlings/propagules
	ii. Interest of locals for plantation
	iii. Manpower and cooperation
	iv. Suitable environment
	v. Capability of withstand risk
	vi. Knowledge of local people
Weakness	i. Cost involvement for plantation
	ii. Lack of enthusiasm of people in participation for long run
Opportunity	i. Maintenance of biodiversity
	ii. Ecosystem services
Threat	i. Frequent natural calamities
	ii. Soil erosion
	iii. Open area for fishing
	iv. Grazing

7.4   STRATEGIC PLANNING

Two viable strategies need to be considered for plantation: (i) Matrix ranking, and (ii) Delegation of responsibility.

7.4.1   MATRIX RANKING

In this strategy, a class of objects is evaluated by different set of criteria and assigning value to the criteria (Table 7.3). This technique reveals the interest of local people to a particular purpose.

Preference of local people includes *Avicennia* species atop in the ranking that necessitates certain purposes like fuel, construction, fodder and timber related to daily domestic need. Establishment of people preference subsequently is to address the ecological restoration for which plantation is undertaken purposefully. For instance, plantation of *Rhizophora* spp., a fast-growing mangrove facilitates shoreline protection by checking soil erosion, along with others to contribute a variety of ecosystem services.

TABLE 7.3 Matrix of Ranking of Mangroves for Local People's Preference (Score 1–10)

Species	Criteria							Total Score	Ranking
	Fire wood	Fodder	Timber	Construction	Thatching	Honey			
Agiceras sp.	04	00	00	04	00	10	18	VIII
Aegialitis sp.	04	00	00	04	00	07	15	X
Aviccenia spp.	10	09	05	07	03	08	42	I
Bruguiera spp.	06	00	04	07	00	00	17	IX
Ceriops spp.	10	00	00	08	00	00	18	VIII
Excoecaria sp.	09	00	00	07	00	05	21	VII
Heritiera spp.	04	00	07	08	00	06	25	V
Kandelia sp.	04	00	00	04	00	03	11	XII
Lumnitzera spp.	04	00	02	04	00	05	15	X
Nypa sp.	02	00	00	00	10	00	12	XI
Phoenix sp.	06	00	04	08	05	00	23	VI
Rhizophora spp.	07	00	10	09	00	00	26	IV
Scyphiphora sp.	04	03	00	03	00	05	15	X
Sonneratia spp.	08	07	06	06	00	05	32	II
Xylocarpus spp.	08	00	08	08	00	05	29	III

7.4.2 DELEGATION OF RESPONSIBILITY

Give responsibility to the respective leader of each group. It is always advisable to involve the local people and arouse their interest in such program because they are the ultimate stakeholders. With involvement of sizable local populations for large-scale plantation, a variety of traditional cultural activities like drama and sacred chanting may be performed that make people enthusiastic about the need of plantation. Sometimes, audiovisual aids may be used to display the benefits of plantation to the people and the society, because no plantation can be successful without participation of local inhabitants (Mandal et al., 2010).

7.5 SILVICULTURE AND SELECTION OF GERMPLASM

Silviculture is the practice of controlling the establishment, growth, composition, health, and quality of forests to meet diverse needs and values (Hawley and Smith, 1954). It also ensures that the treatment of forest stands is used to preserve and make better their productivity. Saenger (2002) suggested that select suitable germplasm first, and then follow propagation for which nursery establishment is essential.

7.5.1 IMPORTANCE FOR ESTABLISHMENT OF MANGROVES NURSERY

The flowering and fruiting periods of most of the mangroves commence during pre-monsoon months and continue till the end of post-monsoon (Table 4.2 in Chapter 4). Biologically, young seedlings/propagules of mangroves have difficulties while growing in high saline condition, because young mangroves do not have organs developed to tolerate high salinity. Rather they prefer growing in low saline condition unless they develop and establish certain organs to excrete/accumulate salt through specific mechanisms as species-specific. Most of the mangroves establish their progeny through recruitment of reproductive units (fruits/propagules) in the period that stretches from peak monsoon (July August) to post monsoon (October-November), an ideal time for regeneration of mangroves. On the contrary, the suitable period for plantation of mangroves propagules/fruits is considered to cover usually pre-monsoon (June) and monsoon (July) months, when fruit maturity does not occur. People interested for plantation have to wait for a period of three months more for planting mangroves by collection of propagules/mature fruits from natural sources. However, such time for plantation seems not to be favorable due to the following reasons: (i) rainfall becomes scanty, (ii) roots of young propagules are not mature enough to anchor substrate, and (iii) salinity gradually increases.

With this backdrop, there is a necessity for raising seedlings in nurseries so that these seedlings can be used for plantation (Figure 7.1 shows raising of seedlings in nursery beds). More and above, nursery raising seedlings will have few advantages for plantation, because:

i)   They have well-developed roots for being raised 8–9 months more in nursery bed, for which they will acclimatize substrate and anchor soil particles easily after plantation.

ii) Segregation of seedlings is undertaken based on their growth so that plantation of seedlings is done as per their suitability to the particular substrate.

iii) There is a scope for weak seedlings to be nurtured one season more till they are suitable for plantation.

iv) There is an option for selection of substrate matching with seedlings as well species preference in relation to health and ability of seedlings to grow.

v)   Loss of seedlings can be checked.

vi) Monitoring and caring becomes easy.

vii) There is a chance to avoid natural hazards to damage seedlings.

viii) It will create employment opportunity to locals.

ix) There is a scope for locals to have interest to learn how to raise mangroves in nursery beds.

7.5.2   COLLECTION REPRODUCTIVE UNITS AND THEIR RAISING IN NURSERY

There may be two approaches for selection of mangroves germplasm: (i) collection of wild seedlings and (ii) raising seedlings in a nursery.

i.   **Collection of wild seedlings**: In established mangroves forests are lying/available an abundant numbers of wild seeds/ seedlings along high tide lines. Collect them and see if they are viable and potential to germinate. Here is listed the details of mangroves reproductive units used for a forestation/plantation/nursery purposes (Table 7.4).

FIGURE 7.1   (a) Seedlings are grown in nursery bed; (b) plantation of young seedlings.

TABLE 7.4 Details of Mangroves Reproductive Units Used for Afforestation/Plantation/Nursery Purposes

Sc. Name	Collection unit	Features of mature unit	Shape	Size of mature unit (cm)	Month for collection	Place for collection	Technique for collection
Acanthus ebracteatus	Fruit	Yellow-brown fruit coat	Small cylindrical	1.5–2.0	July–Oct.	Around the mother plant	Manual
Acanthus ilicifolius	Fruit	Yellow-brown fruit coat	Small cylindrical	2.0–2.5	July–Oct.	Around the mother plant	Manual
Aegialitis rotundifolia	Fruit	Rupture fruit coat with cotyledon and floating ability	Round rod	3.0–4.0	Aug.–Nov.	Under mother tree/ far from mother tree	Manual/drag net
Aegiceras corniculatum	Fruit	Rupture fruit coat with cotyledon and floating ability	Round rod with pointed end	3.0–4.0	Aug.–Nov.	Under mother tree/ far from mother tree	Manual/drag net
Avicennia alba	Fruit	Cotyledon with rupture fruit coat and floating ability	Conical	2.0–4.0	Aug.–Nov.	Under mother tree/ far from mother tree	Manual/drag net
Avicennia marina	Fruit	Cotyledon with rupture fruit coat and floating ability	Flattened with short apical beak	1.2–2.5	Aug.–Nov.	Under mother tree/ far from mother tree	Manual/drag net
Avicennia officinalis	Fruit	Cotyledon with rupture fruit coat and floating ability	Flattened spherical	0.28–3.0	Aug.–Nov.	Under mother tree/ far from mother tree	Manual/drag net
Bruguiera cylindrica	Propagule	Greenish hypocotyl with floating ability	Round rod	14.0–16.0	Nov. –Feb.	Under mother tree/ far from mother tree	Manual/drag net
Bruguiera gymnorhiza	Propagule	Reddish-green hypocotyl with floating ability	Round rod	16.0–18.0	Nov. –Feb.	Under mother tree/ far from mother tree	Manual/drag net
Bruguiera parviflora	Propagule	Yellowish-green hypocotyl with floating ability	Round long rod	20.0–25.0	Nov. –Feb.	Under mother tree/ far from mother tree	Manual/drag net
Bruguiera sexangula	Propagule	Yellowish-green hypocotyl with floating ability	Round rod	8.0–12.0	Nov. –Feb.	Under mother tree/ far from mother tree	Manual/drag net

TABLE 7.4 (Continued)

Sc. Name	Collection unit	Features of mature unit	Shape	Size of mature unit (cm)	Month for collection	Place for collection	Technique for collection
Ceriops decandra	Propagule	Yellowish-green hypocotyl with floating ability	Round rod	8.0–12.0	Oct.–Dec.	Under mother tree/far from mother tree	Manual/drag net
Ceriops tagal	Propagule	Greenish hypocotyl with floating ability	Round long rod	20.0–25.0	Oct.–Dec.	Under mother tree/far from mother tree	Manual/drag net
Excoecaria agallocha	Seed	Small, numerous with germinated sprout	Round		July–Aug.	Under mother tree	Manual
Heritiera fomes	Fruit	Grayish with short keel and floating ability	Almost round	1.8–2.0	Aug.–Oct.	Under mother tree/far from mother tree	Manual/drag net
Heritiera littoralis	Fruit	Brownish with long keel and floating ability	Flattened long	5.0–6.0	Oct.–Dec.	Under mother tree/far from mother tree	Manual/drag net
Kandelia candel	Propagule	Lush green hypocotyl with floating ability	Long rod	40.0–45.0	July–Sept.	Under mother tree/far from mother tree	Manual/drag net
Lumnitzera littorea	Fruit	Greenish	Flattened elliptic	1.2–1.5	Oct.–Dec.	Under mother tree	Manual
Lumnitzera racemosa	Fruit	Greenish	Flattened elliptic	0.9–1.0	Aug.–Nov.	Under mother tree	Manual
Nypa fruticans	Fruit	Crypto-viviparous	Round	10.0–12.0	July–Sept.	Under mother tree	Manual
Rhizophora apiculata	Propagule	Yellowish-green hypocotyl with floating ability	Long rod	40–55	Aug.–Oct.	Under three/far from mother tree	Manual/drag net
Rhizophora mucronata	Propagule	Yellowish-reddish hypocotyl with floating ability	Long rod	50–75	Aug.–Oct.	Under three/far from mother tree	Manual/drag net

TABLE 7.4 *(Continued)*

Sc. Name	Collection unit	Features of mature unit	Shape	Size of mature unit (cm)	Month for collection	Place for collection	Technique for collection
Rhizophora stylosa	Propagule	Yellowish-brown hypocotyl with floating ability	Medium rod	20–40	Aug.–Oct.	Under three/far from mother tree	Manual/drag net
Scyphiphora hydrophylacea	Fruit	Greenish	Elliptic	1.0–1.2	July–Sept.	Under mother tree	Manual
Sonneratia alba	Fruit	Greenish	Round and flattened	4.5–7.5	June–Sept.	Under mother tree	Manual
Sonneratia apetala	Fruit	Greenish	Round	1.5–2.0	July–Sept.	Under mother tree	Manual
Sonneratia caseolaris	Fruit	Greenish	Round and flattened	5.0–8.0	June–Oct.	Under mother tree	Manual
Sonneratia griffithii	Fruit	Greenish	Round and flattened	5.0–7.0	June–Sept.	Under mother tree	Manual
Sonneratia ovata	Fruit	Greenish	Round and flattened	4.0–6.0	July–Sept.	Under mother tree	Manual
Xylocarpus granatum	Fruit/seed	Brown with floating ability	Round	> 15	July–Sept.	Under three/far from mother tree	Manual/drag net
Xylocarpus moluccensis	Fruit/seed	Greenish with floating ability	Round	< 10	July–Nov.	Under three/far from mother tree	Manual/drag net

ii.   **Raising seedlings in a nursery**: Collect wild seeds from the forest and grow them in a nursery bed, not difficult for preparation and seeds with minimum care may germinate to develop seedlings. There are some books exclusively dealing with silviculture practice.

7.5.3   CRITERIA FOR SUITABLE SITE FOR MANGROVES GROWTH

There are few criteria which need to be considered for survival and growth of mangroves. We here mention the following criteria suitable for growth of mangroves seedlings.

(a)   Salinity ranges of substrate, the basis of mangroves tolerance and survival, are important for seedlings survival. Based on the salinity range, there are five zones: (i) Euhaline zone (> 30‰), Polyhaline zone (18 to 30‰), Mesohaline zone (5 to 18‰), Oligohaline zone (0.5 to 5‰) and Limnatic condition (<0.5‰). Most of the mangroves seedlings can easily grow in Oligohaline zone, and may also survive in Mesohaline zone subject to plantation of healthy seedlings with well-developed roots. However, not all mangroves will survive in this zone, only those mangroves which tolerate and grow in Polyhaline zone may survive (Figure 7.2).

(b)   Substratum should be muddy or have deposit of accumulated sediments in association with drainage systems including permanent rivers and creeks that provide freshwater and sediment regularly or with regular intervals. Because sediment rich environments will facilitate sediment redistribution and accretion.

(c)   Soil needs to be saturated with freshwater or covered by low saline water at some time during the growing season of mangroves seedlings or flooded regularly with low saline tidal water until seedlings get established.

7.5.4   CRITERIA FOR UNSUITABLE SITE FOR MANGROVES GROWTH

(a)   Sandy substrate are not suitable for mangroves plantation due to the following reasons: (i) moving sands are likely to be deposited on leaves and damage young seedlings, (ii) mangroves do not have deep rooting so that they may not grip properly sandy soil, (iii) sandy substrate have less water retention capability and low quantity of nutrient contents, (iv) sandy soil are not a suitable substrate for sediment deposition, and (v) sandy soil is characteristic with loose particles that facilitate erosion by which mangroves seedlings are subjected to be vulnerable for uprooting.

(b) Plantation of seedlings should be avoided in such substrate which are highly infested with a variety of crabs and mollusks which may likely graze on mangroves seedlings and thus damage their growth and establishment.

7.6 SITE SELECTION

Human activities, by and large, are responsible for damaging or destroying mangrove vegetation. There are many areas where mangrove rehabilitation may be attempted, which include areas for developing shrimp ponds by clearing of mangrove, areas used for charcoal production and brick industry development, or even areas in which mangroves drying out as a result of nearby changes in hydrology (due to construction of dikes, levees, roads, upland deforestation). In these places, before planting mangroves it is imperative to determine if the areas are presently suitable for mangrove growth, with consultation of local community. Otherwise alternative areas need to be selected and protected from the adversities of strong tidal thrust, frequent soil erosion and be fairly non-polluted zone. All these adversities are likely to disturb the normal growth of seedlings. This survey helps the planners to explore suitable places in which plantation may be undertaken. Here are mentioned few points to achieve successful mangrove restoration following Lewis and Marshall (1997);

i.   Understand the autecology (individual species ecology) of the mangrove species, particularly the patterns of reproduction, phenology, propagule distribution, and successful seedling establishment (see Chapter 4).

ii. Understand the normal hydrologic patterns that control the distribution and successful establishment and growth of targeted mangrove species (see Chapter 3).

iii. Assess modifications of the original mangrove environment.

If the understanding of site selection for mangroves plantation is fulfilled, other two purposes such as planting technique, followed by monitoring are essentially required for successful plantation.

7.6.1 PLANTING TECHNIQUES

- Broadcast fruits/propagules directly on the surface.
- Prepare a number of holes with spacing between 1.5×1.5 m^2 suitable for planting.
- Follow the pattern of zonation, along with species combination (Figure 7.2).
- Transplant wild seedlings or transplant raising seedlings collected from a nursery bed.
- For viviparous propagules, segmented parts may be planted (Saenger, 2002).
- Fill up hole with loose soil so that aerial spaces remain sufficient for seedlings to survive.

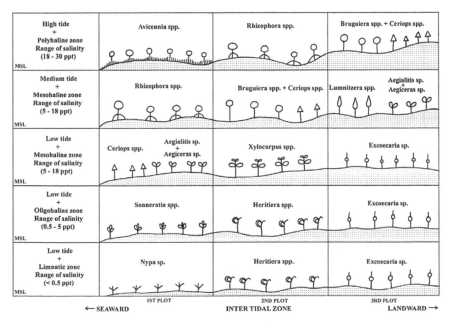

FIGURE 7.2   Plantation of mangrove seedlings based on saline zones.

7.6.2 MONITORING

Post plantation program is probably the toughest job that needs concerted effort from all the partners involved in restoration program. Monitoring requires in regular intervals that measure the growth of seedlings as a function of time. A growth characteristic that determines the progress of plantation includes some parameters like height, stem structure, node production, phenology, fruiting and resistance to pests of individuals, besides density, percent cover and species composition over time.

7.7 CONSERVATION

Depletion of mangroves has been an evident worldwide over a period of time. A concerted effort is essentially required to conserve them. This book is to focus some points which are supposed to be regional specific, but address few attempts to protect and conserve mangroves.

i.   **Awareness campaign**: A curriculum in school syllabus needs to be included to aware students about the importance of mangroves in protecting coastal areas because coastal regions are densely populated, around 90% population lives within 100 km of the coast in the world (Saenger, 2002). Their participation in protecting mangroves may be undertaken through education and is likely to be effective. Even, students of college and universities need to have compulsory participation for plantation activities and their performance to be assessed based on how many seedlings one student is to plant, protect and develop. Though such activity may be convenient to the students inhabiting near coastal areas.

ii. **Restriction of grazing**: Some areas are used for grazing indiscriminately. However, once the areas are used for mangroves plantation, they are declared as prohibited zones from doing all such activates.

iii. **Restriction of waste runoff**: Most of the urban sewage effluent are discharged and runoff through mangroves habitats. This is a serious problem that is one of the causative factors to destroy mangroves.

iv. **Imposition of legislation**: A set of laws may be imposed to protect mangroves habitats. It is a Government initiative, but the legislation may be introduced slowly to assess how much it may be effective. There are many laws already imposed, but execution of laws seems to be poor. It may be considered one among other ways to protect already vulnerable mangroves forest.

v. **Alternative livelihoods**: Most of the people living in coastal areas are poor and fully dependent on collecting resources from mangroves forest and surrounding waterways since mangroves ecosystem are highly productive system. Even when the lean period of normal agriculture emerge, government needs to cater foods and other necessary items with subsidized rates so that their dependence on forest may be reduced.

7.8   MANGROVES ECOLOGICAL PARK

In Indian Sundarban, an ecological park has been established covering a large-scale area at Jharkhali. Dr. K. R. Naskar, an eminent mangroves worker conceived the very idea to showcase all the mangroves species to be available in one place. With this intention, his team collected wild seedlings from other parts of Indian Sundarban and planted them in consideration of hydrology and zonation of respective species, with other existing species. Dr. Naskar religiously monitored their growth with scientific spirit over a period of 10 years. After its successful establishment, it was transferred to forest department on request. Presently the ecological park maintains a variety of mangroves species in one place, which are distributed in different parts of the forest. The idea of Ecological Park now benefits schools children, college and University students who enjoy seeing all the species of mangroves, their morphology, shapes, diversity and other dependents fauna upon them. Now, this is a place of ecotourism.

7.9   KNOW YOUR MANGROVES

Human resources, particularly those residing around mangroves areas, are considered the ultimate stakeholders for protection of mangroves. They are of diverse communities in culture with regional specific throughout the globe, but their way of using mangroves appears to be similar across the global boundary. Presently the importance of mangroves invites special attention to the environmentalists in the context of climate change, global warming and sea level rise that indicate an immediate threat as submergence of coastal low land areas in the globe. To mitigate such threat, the role of mangroves vegetation is felt noticeable. Already there is a caveat that if the destruction of mangroves continues, this unique coastal vegetation disappears from this world within 100 years (Duke et al., 2007). Before the damage occurs, we need to act displaying the banner 'Know Your Mangroves and Protect Them;' else we certainly miss the wonderful creatures of nature.

KEYWORDS

- conservation
- mangroves ecological park
- site selection
- strategic planning

BIBLIOGRAPHY

Adegbehin, J. O. *Mangroves in Nigeria*. In *Conservation and Sustainable Utilization of Mangroves Forests in Latin America and African Regions (Part 2: Africa). Mangroves Ecosystem Technical Reports 3*. Diop, E. S., (Ed.). International Society for Mangroves Ecosystems and International Tropical Timber Organization: Tokyo, 1993; pp. 1–153.

Allen, J. R. Constraints on measurement of sea level movements from salt-marsh accretion rates. *J. Geol. Soc. Lond.* **1990,** *147,* 5–7.

Allen, J. R. *Tidally induced marshes in the Severn Estuary, southwest Britain*. In *Saltmarshes: Morphodynamics, Conservation and Engineering Significance.*Allen, J. R.; Pye, K., (Eds.). Cambridge University Press: Cambridge, 1992; p. 123–147.

Alongi, D. M. Mangrove forests: Resilience, protection from tsunamis, and responses to global climate change. *Estuarine, Coastal and Shelf Science.* **2008,** *76,* 1–13.

Alongi, D. M. *Mangrove-microbe-soil relations*. In *Interactions between macro and micro organisms in marine sediments;* Kristensen, E.; Haese, R. R.; Kostka, J. E. Eds.; American Geophysical Union: Washington, D.C., **2005,** pp. 85–103.

Alongi, D. M. Present state and future of the world's mangrove forests. *Environ. Conserv.* **2002,** *29,* 331–349.

Alongi, D. M. The Impact of Climate Change on Mangrove Forests. *Curr. Clim. Change. Rep.;* **2015,** *1,* 30–39.

Andrews, T. J.; Clough, B. F.; Muller, G. J. *Photosynthetic gas exchange properties and carbon isotope ratios of some mangroves in North Queens-land*. In: *Physiology and Management of Mangroves, Tasks for Vegetation;* Teas, H. J. Ed.;Science, Dr. W. Junk: The Hague, 1984; Vol. 9; p 15–23.

Auerbach, L. W.; Goodbred, S. L. Jr.; Mondal, D. R.; Wilson, C. A.; Ahmed, K. R. Flood risk of natural and embanked landscapes on the Ganges–Brahmaputra tidal delta plain. *Nat. Clim. Change.* **2015,** *5,* 153–157.

Bagarinao, T. U. Systematics, genetics, distribution and life history of milkfish Chanoschanos. *Environ. Biol. Fish.* **1994,** *39,* 23–41.

Baldwin, A.; Egnotovich, M.; Ford, M.; Platt, W. Regeneration in fringe mangrove forests damaged by Hurricane Andrew. *Plant Ecol.* **2001,** *157,* 149–162.

Ball, M. C.; Cochrane, M. J.; Rawason, H. M. Growth and water use of the mangroves *Rhizophora apiculata* and *R. stylosa* in response to salinity and humidity under ambient and elevated concentration of atmospheric CO_2. *Plant Cell Environ.* **1997,** *20,* 1158–1166.

Ball, M. C.; Munns, R. Plant responses to salinity under elevated atmospheric concentrations of CO_2. *Australian Journal of Botany.* **1992,** *40,* 515–525.

Bandaranayake, W. M. Traditional and medicinal uses of mangroves. *Mangroves and Salt Marshes.* **1998,** *2*(3), 133–148.

Barber, C. V.; Miller, K. R.; Boness, M. *Securing protected areas in the face of global change: issues and strategies*. Gland, Switzerland and Cambridge, UK: International Union for Conservation of Nature and Natural Resources. 2004.

Barletta-Bergan, A.; Barletta, M.; Saint-Paul, U. Community structure and temporal variability of ichthyoplankton in North Brazilian mangrove creeks. *J. Fish. Biol.* **2002,** *61,* 33–51.

Basit, M. A. *Non-wood forest products from the mangroves forests of Bangladesh.* In *Beyond Timber: Social, Economic and Cultural Dimensions of Non Wood Forest Products.* Food and Agriculture Organization of the United Nations: Rome, Report 13, 1995.

Bennett, E. M.; Cumming, G. S.; Peterson, G. D. A systems model approach to determining resilience surrogates for case studies. *Ecosystems.* **2005,** *8,* 945–957.

Benzie, J. A.H. Genetic structure of coral reef organisms, ghosts of dispersal past. *Am. Zool.* **1999,** *39,* 131–145.

Berdin, R.; Siringan, F.; Maeda, Y. Holocene relative sea-level changes and mangrove response in Southwest Bohol, Philippines. *J. Coast. Res.* **2003,** *19,* 304–313.

Bindoff, N. L.; Willebrand, J.; Artale, V.; Cazenave, A.; Gregory, J.; Gulev, S.; Hanawa, K.; Le Que´re´, C.; Levitus, S.; Nojiri, Y.; Shum, C.; Talley, L.; Unnikrishnan, A. *Observations: oceanic climate change and sea level.* In *Climate Change 2007:The Physical Science Basis. Contribution of Working Group I to the Fourth Assessment Report of the Intergovernmental Panel on Climate Change.* Solomon, S.; Qin, D.; Manning, M.; Chen, Z.; Marquis, M.; Averyt, K.; Tignor, M.; Miller, H. Eds.; Cambridge University Press: Cambridge, United Kingdom and New York, NY, USA, 2007.

Blasco, F. *Taxonomic considerations of mangrove species.* In *The mangrove ecosystem: research methods;* Snedaker, S. C.; Snedaker, J. G., (Eds.). UNESCO: Paris, 1984.

Blasco, F.; Aizpuru, F. Mangroves along the coastal stretch of the Bay of Bengal: present status. *Indian J. Marine Sci.* **2002,** *31,* 9–20.

Blasco, F.; Saenger, P. Janodet, E. Mangroves as indicators of coastal change. *Catena.* **1996,** *27,* 167–178.

Boesch, D. F. *Diversity, stability and response to human disturbance in estuarine ecosystems.* In *Proceedings of the First International Congress of Ecology,* Wageningen: The Netherlands, **1974,** p. 109–114.

Brander, M.; Wagtendonk, A. J.; Hussain, S. S.; Vittie, A. Mc.; Verburg, P. H.; Groot, R. S.; Ploeg, S. van der. Ecosystem service values for mangroves in Southeast Asia: A meta-analysis and value transfer application Luke. *Ecosystem Services,* **2012,** *1,* 62–69.

Burchett, M. D.; Meredith, S.; Pulkownick, A.; Walsh, S. Short term influences affecting growth and distribution of mangrove communities in the Sydney region. *Wetlands.* **1984,** *4,* 63–72.

Cabanes, C.; Cazenave, A.; Le Provost, C. Sea-level rise during the past 40 years determined from satellite and in situ observations. *Science.* **2001,** *294,* 840–842.

Cahoon, D. R. A review of major storm impacts on coastal wetland elevations. *Estuar. Coast.* **2006,** *29,* 889–898.

Cahoon, D. R.; Day, J. W.; Reed, D. J. The influence of surface and shallow subsurface soil processes on wetland elevation, a synthesis. *Curr. Topics Wetland Biogeochem.* **1999,** *3,* 72–88.

Cahoon, D. R.; Guntenspergen, G. R. Climate change, sea-level rise, and coastal wetlands. *Nat. Wetl. Newslett.* **2010,** *32,* 8–12.

Cahoon, D. R.; Hensel, P. F.; Spencer, T.; Reed, D. J.; McKee, K. L.; Saintilan, N. *Coastal wetland vulnerability to relative sea-level rise: wetland elevation trends and process controls.* In *Wetlands and Natural Resource Management; Ecological Studies;* Vol. 190;

Verhoeven, J. T. A.; Beltman, B.; Bobbink, R.; Whigham, D. Eds.; Springer-Verlag: Berlin/Heidelberg, 2006; pp. 271–292.

Cahoon, D. R.; Hensel, P. In Proceedings of the Symposium on Mangrove Responses to Relative Sea Level Rise and Other Climate Change Effects; Catchments to Coast, Society of Wetland Scientists 27th International Conference, July 9–14,2006,Gilman, E. Ed.; Cairns Convention Centre: Cairns, Australia. Western Pacific Regional Fishery Management Council, Honolulu, HI, USA, ISBN: 1-34061-03-4, 2006; pp. 9–17.

Cahoon, D. R.; Hensel, P.; Rybczyk, J.; McKee, K.; Proffitt, C. E.; Perez, B. Mass tree mortality leads to mangrove peat collapse at Bay Islands, Honduras after Hurricane Mitch. *J. Ecol.* **2003,** *91,* 1093–1105.

Cahoon, D. R.; Hensel, P.; Rybczyk, J.; Perez. B. C. Hurricane Mitch: impacts on mangrove sediment elevation dynamics and long-term mangrove sustainability: USGS Open File Report 03–184, p. 75, 2002c.

Cahoon, D. R.; Lynch, J. C. Vertical accretion and shallow subsidence in a mangrove forest of southwestern Florida, USA. *Mangroves Salt Marshes.* **1997,** *1,* 173–186.

Cahoon, D. R.; Lynch, J. C.; Hensel, P.; Boumans, R.; Perez, B. C.; Segura, B.; Day Jr.; J. W. A device for high precision measurement of wetland sediment elevation: I. Recent improvements to the sedimentation-erosion table. *J. Sediment. Res.* **2002a,** *72,* 730–733.

Callaway, J. C.; Cahoon, D. R.; Lynch, J. C. In *Methods in Biogeochemistry of Wetlands;* De Laune, R. D. et al. Eds.; Soil Science Society of America: **2013,** 901–917.

Cannicci, S.; Burrows, D.; Fratini, S.; Smith III, T. J.; Offenberg, J.; Dahdouh- Guebas, F. Faunistic impact on vegetation structure and ecosystem function in mangrove forests: A review. *Aquat. Bot.* **2008,** *89,* 186–200.

Carpenter, S.; Walker, B.; Anderies, J. M.; Abel, N. From metaphor to measurement: resilience of what to what? *Ecosystems.* **2001,** *4,* 765–781.

Catalogue of Life website. http://www.catalogueoflife.org (accessed on 11 Jul 2018).

Chapman, V. J. Introduction. In *Ecosystems of the world. 1. Wet Coastal Ecosystems;* Chapman, V. J., (Ed.). Elsevier Sci. Publ. Co: Amsterdam, **1977,** *6.*

Chapman, V. J. Mangrove Vegetation; J. Cramer: Germany, 1976.

Chinese Academy of Sciences (CAS). **2011,** *8, 22*http://news.xinhuanet.com/english2010/china/2011–08/22/c_131067137.htm.(Retrieved2017–17–02.)

Chopra, R. N.; Naayar, S. L.; Chopra, I. C. *Glossary of Indian Medicinal plants*, CSIR: New Delhi, 1986.

Christensen, B. Management and Utilization of Mangroves in Asia and the Pacific. FAO Environment Paper No. 3. Food and Agriculture Organization: Rome, 1982.

Christopher, W. S.;White, N. J.; Church, J. A.; King, M. A.; Burgette, R. J.; Legresy, B.Unabated global mean sea level rise over the satellite altimeter era. *Nature Climate Change.* **2015,** *5,* 565–568,doi:10.1038/nclimate2635.

Church, J. A.; Clark, P. U.; Cazenave, A.; Gregory, J. M.; Jevrejeva, S.; Levermann, A.; Merrifield, M. A.; Milne, G. A.; Nerem, R. S.; Nunn, P. D.; Payne, A. J.; Pfeffer, W. T.; Stammer, D.; Unnikrishnan, A. S. *Sea Level Change.* In *Climate Change 2013;The Physical Science Basis. Contribution of Working Group I to the Fifth Assessment Report of the Intergovernmental Panel on Climate Change;*Stocker, T. F.; Qin, D.; Plattner, G.-K.; Tignor, M.; Allen, S. K.; Boschung, J.; Nauels, A.; Xia, Y.; Bex, V.; Midgley, P. M., (Eds.). Cambridge University Press: Cambridge, United Kingdom and New York, NY, USA, 2013.

Church, J. A.; White, N. J. Sea-level rise from the late 19th to the early 21st century. *Surveys in Geophysics.* **2011,** 32 (4–5), 585–602.

Church, J.; Gregory, J.; Huybrechts, P.; Kuhn, M.; Lambeck, K.; Nhuan, M.; Qin, D.; Woodworth, P. *Changes in sea level.* In *Climate Change 2001: The Scientific Basis* (Published for the Intergovernmental Panel on Climate Change); Houghton, J.; Ding, Y.; Griggs, D.; Noguer, M.; van der Linden, P.; Dai, X.; Maskell, K.; Johnson, C. Eds.; Cambridge University Press: Cambridge, United Kingdom, and New York, NY, USA, 2001; pp. 639–693.

Church, J.; Hunter, J.; McInnes, K.; White, N. In Coast to Coast '04—Conference Proceedings Australia's National Coastal Conference, Hobart, April 19–23, **2004b,** p. 1–8.

Church, J.; White, N. A 20th century acceleration in global sea-level rise. *Geophys. Res. Lett.* **2006,** *33,* L01602.

Church, J.; White, N.; Coleman, R.; Lambeck, K.; Mitrovica, J. Estimates of the regional distribution of sea-level rise over the 1950 to 2000 period. *J. Climate.* **2004a,** *17,* 2609–2625.

Cintron, G.; Lugo, A. E.; Martinez, R. *Structural and functional properties of mangroveforests.* In *The botany and Natural History of Panama*; Darcy, W. G.; Correa, M. D., (Eds.). *Monogr.Syst. Bot.* Missouri Bot. Gard.: St. Louis, **1985,** *10,* 53–66.

Clough, B. F. *Mangroves.* In *Control of Crop Productivity.* Academic Press: Sydney, **1984,** pp. 253–268.

Clough, B. F.; Andrews, T. J.; Cowan, I. R. *Physiological processes in mangroves.* In *Mangrove Ecosystems in Australia: Structure, Function and Management.* Clough, B. F. Ed.; Australian National University Press: Canberra, 1982; pp. 193–210.

Cohen, J. E.; Small, C.; Mellinger, A.; Gallup, J.; Sachs, J. Estimates of coastal populations. *Science* **1997,** *278,* 1209–1213.

Costanza, R.; Groot, R.de.; Sutton, P.; Ploeg, S. van der.; Anderson, S. J.; Kubiszewski, I.; Farber, S.; Turner, R. K. Changes in the global value of ecosystem services. *Global Environmental Change,* **2014,** *26,* 152–158.

Crona, B. I.; Ronnback, P. Community structure and temporal variability of juvenile fish assemblages in natural and replanted mangroves, *Sonneratia alba* Sm.; of Gazi Bay, Kenya. *Estuar. Coast. Shelf Sci.* **2007,** *74,* 44–52.

Dahdouh-Guebas, F.; Hettiarachchi, S.; Lo Seen, D.; Batelaan, O.; Sooriyar- achchi, S.; Jayatissa, L. P.; Koedam, N. Transitions in ancient inland freshwater resource management in Sri Lanka affect biota and human populations in and around coastal lagoons. *Curr. Biol.* **2005b,** *15,* 579–586.

Dahdouh-Guebas, F.; Jayatissa, L. P.; Di Nitto, D.; Bosire, J. O.; LoSeen, D.; Koedam, N. How effective were mangroves as a defence against the recent tsunami? *Curr. Biol.* **2005a,** *15,* R443–R447.

Dahdouh-Guebas, F.; Koedam, N. Long-term retrospection on mangrove development using transdisciplinary approaches: A review. *Aquat. Bot.* **2008,** *89,* 80–92.

Dahdouh-Guebas, F.; Koedam, N.; Danielsen, F.; Sørensen, M. K.; Olwig, M. F.; Selvam, V.; Parish, F.; Burgess, N. D.; Topp-Jørgensen, E.; Hiraishi, T.; Karunagaran, V. M.; Rasmussen, M. S.; Hansen, L. B.; Quarto, A.; Suryadiputra, N. Coastal vegetation and the Asian tsunami. *Science.* **2006,** *311,* 37– 38.

Danielsen, F.; Soerensen, M.; Olwig, M.; Selvam, V.; Parish, F.; Burgess, N.; Hiraishi, T.; Karunagaran, V.; Rasmussen, M.; Hansen, L.; Quarto, A.; Nyoman, S. The Asian tsunami: a protective role for coastal vegetation. *Science.* **2005,** *310,* 643.

Das, C.S; Mandal R. N. Coastal people and mangroves ecosystem resources vis-à-vis management strategies in Indian Sundarban. *Ocean & Coastal Management.* **2016,** 134, 1–10.

Ding, Hou. Rhizophoraceae. *Flora Malesiana.* Series 1, **1958,** *5,* 429–493.

Diop, E. S.; Ed. Conservation and Sustainable Utilization of Mangroves Forests in Latin America and African Regions (Part 2: Africa). Mangroves Ecosystem Technical Reports 3. International Society for Mangroves Ecosystems and International Tropical Timber Organization, Tokyo, 1993.

Dittmar, T.; Hertkorn, N.; Kattner, G.; Lara R. J. Mangroves, a major source of dissolved organic carbon to the oceans. *Global biogeochem cycles.* **2006,** 20: GB1012. Doi:10.1029/2005GB002570.

Dodd, R. S.; Ong, J. E. *Future of mangrove ecosystems to 2025.* In *Aquatic ecosystems: Trends and global propects;* Polunin, N. V. C., (Ed.). Cambridge: Foundation of Environmental Conservation, Cambridge university Press: Cambridge, 2008.

Drexler, J. Z.; Ewel, K. C. Effect of the 1997–1998 ENSO-Related Drought on hydrology and salinity in a Micronesian wetland complex. *Estuaries.* **2001,** *24,* 347–356.

Duke, N. C. *Mangrove floristics and biogeography.* In *Tropical Mangrove Ecosystems;* Robertson, A. I.; Alongi, D. M. Eds.; American Geophysical Union: Washington, DC, USA, 1992; pp. 63–100.

Duke, N. C.; Ball, M. C.; Ellison, J. C. Factors influencing biodiversity and distributional gradients in mangroves. *Global Ecol. Biogeogr. Lett.*1998, *7,* 27–47.

Duke, N. C.; Meynecke, J. O.; Dittmann, S.; Ellison, A. M.; Anger, K.; Berger, U.; Cannicci, S.; Diele, K.; Ewel, K. C.; Field, C. D.; Koedam, N.; Lee, S. Y.; Marchand, C.; Nordhaus, I.; Dahdouh-Guebas, F. A world without mangroves? *Science.* **2007,** *317,* 41–42.

Edwards, A. *Impact of climate change on coral reefs, mangroves, and tropical seagrass ecosystems.* In *Climate Change Impact on Coastal Habitation;* Eisma, D.; Ed. Lewis Publishers: 1995.

Elizabeth, M.; Salm, R. V. Managing Mangroves for Resilience to Climate Change. IUCN: Gland, Switzerland, **2006,** p.64.

Ellison, A. M. Managing mangroves with benthic biodiversity in mind: moving beyond roving banditry. *J. Sea Res.* **2008,** *59,* 2–15.

Ellison, A. M. Mangrove restoration: Do we know enough? *Restoration Ecology;* **2000,** *8*(3), 219.

Ellison, A. M.; Farnsworth, E. J. Mangrove communities. In *Marine Community Ecology;* Bertness, M. D.; Gaines, S. D.; Hay, M. E. Eds.; Sinauer Associates: New York, 2000; pp. 423–442.

Ellison, A. M.; Farnsworth, E. J.; Merkt, R. E. Origins of mangrove ecosystems and mangrove biodiversity anomaly. *Global Ecol. Biogeogr. Lett.* **1999,** *8,* 95–115.

Ellison, J. C. *How South Pacific mangroves may respond to predicted climate change and sea-level rise.* In *Climate Change in the South Pacific: Impacts and Responses in Australia, New Zealand, and Small Islands States;* Gillespie, A.; Burns, W., (Eds.). Kluwer Academic Publishers: Dordrecht, 2000; pp. 289–301.

Ellison, J. C. Impacts on mangrove ecosystems. The Great Greenhouse Gamble: A conference on the Impacts of Climate Change on Biodiversity and Natural Resource Management: Conference Proceedings, Sydney, NSW, EJ. 2005.

Ellison, J. C. Long-term retrospection on mangrove development using sediment cores and pollen analysis: A review. *Aquat. Bot.* **2008,** *89,* 93–104.

Ellison, J. C. Mangrove retreat with rising sea level, Bermuda. *Estuarine Coastal and Shelf Science.* **1993,** *37,* 75–87.

Ellison, J. C.; Stoddart, D. R. Mangrove ecosystem collapse during predicted sea-level rise: Holocene analogues and implications. *Journal of Coastal Research.* 1991; 7, 151–165.

Ellison, J. C. Vulnerability of Fiji's mangroves and associated coral reefs to climate change. Review for the World Wildlife Fund. Launceston, Australia: University of Tasmania; 2004.

Ellison, J. *Mangrove paleoenvironmental response to climate change.* In Proceedings of the Symposium on Mangrove Responses to Relative Sea-Level Rise and Other Climate Change Effects, Society of Wetland Scientists 2006 Conference, 9–14 July2006,Cairns, Australia. Western Pacific Regional Fishery Management Council and United Nations Environment Programme Regional Seas Programme, Honolulu, Gilman, E. Eds; USA and Nairobi, Kenya, ISBN: 1-934061-03-4; 2006; pp. 1–8.

Ellison, J. Mangrove retreat with rising sea level, Bermuda. *Estuarine Coastal Shelf Science,* **1993,** *37,* 75–87.

Ewel, K. C.; Twilley, R. R.; Ong, J. E. Different kinds of mangrove forests provide different goods and services. *Global Ecol. Biogeogr. Lett.*1998, *7,* 83–94.

FAO (Food and Agriculture Organization of the United Nations). The World's mangroves 1980–2005. FAO Forestry Paper 153, FAO, Rome, 2007.

FAO. Mangroves Forest Management Guidelines. FAO Forestry Paper 117. Food and Agriculture Organization of the United Nations, Rome, 1994.

FAO. Mangroves Management in Thailand, Malaysia and Indonesia. FAO Environment Paper 4. Food and Agriculture Organization of the United Nations, Rome, 1985.

Farnsworth, E. J.; Ellison, A. M.; Gong, W. K. Elevated CO_2 alters anatomy, physiology, growth and reproduction of red mangrove (*Rhizophoramangle* L.). *Oecologia.* **1996,** *108,* 599–609.

Field, C. D.; Hinwood, B. G.; Stevenson, I. *Structural features of salt gland of Agiceras.* In *Physiology and management of mangroves*; Teas H. J., (Ed.). Vol. 8. Dr. W. Junk, The Hague, 1984.

Field, C. B.; Osborn, J. G.; Hoffman, I. J.;Ackerly D. D. Mangrove biodiversity and ecosystem function. *Global Ecol. Biogeogr. Lett.*1998, *7,* 3–14.

Field, C. Impacts of expected climate change on mangroves. *Hydrobiologia.*1995, *295,* 75–81.

Fischlin, A.; Midgley, G. F.; Price, J. T.; Leemans, R.; Gopal, B.; Turley, C.; Rounsevell, M. D. A.; Dube, O. P.; Tarazona, J.; Velichko, A. A. *Ecosystems, their properties, goods and services.* In Climate change **2007,** Impacts, Adaptation and Vulnerability. Contribution of Working Group II to the Fourth Assessment Report of the Intergovernmental Panel of Climate Change (IPCC); Parry, M. L.; Canziani, O. F.; Palutikof, J. P.; van der Linden, P. J.; Hanson, C. E., (Eds.). Cambridge University Press: Cambridge, UK, 2007; pp. 211–272.

French, J. R. *Eustatic and neotectonic controls on salt marsh sedimentation.* In *Coastal Sediments '91.* Kraus, N. C.; Gingerich, K. J.; Kriebel, D. L. Eds.; American Society of Civil Engineers: New York, 1991; pp, 1223–1236.

French, J. R. Numerical simulation of vertical marsh growth and adjustment to accelerated sea-level rise, North Norfolk, UK. *Earth Surf. Proc. Land.* **1993,** *18,* 63–81.

Gilman, E.; Ellison, J.; Coleman, R. Assessment of mangrove response to projected relative sea-level rise and recent historical reconstruction of shoreline position. *Environ. Monit. Assess.* **2007,** *124,* 112–134.

Gilman, E.; Ellison, J.; Duke, N. C.; Field, C. Threats to mangroves from climate change and adaptation options: A review. *Aquatic Botany.* **2008,** *89,* 237–250.

Gilman, E.; Ellison, J.; Jungblat, V.; VanLavieren, H.; Wilson, L.; Areki, F.; Brighouse, G.; Bungitak, J.; Dus, E.; Henry, M.; Sauni Jr.; I.; Kilman, M.; Matthews, E.; Teariki-Ruatu, N.; Tukia, S.; Yuknavage, K. Adapting to Pacific Island mangrove responses to sea level rise and other climate change effects. *Climate Res.* **2006,** *32,* 161–176.

Gilman, E.; Ellison, J.; Sauni Jr.; I.; Tuaumu, S. Trends in surface elevations of American Samoa mangroves. *Wetl. Ecol. Manag.* **2007,** *15,* 391–404.

Gilmore, R. G.; Snedaker, S. C. Mangrove forests. In *Biodiversity of the Southeastern United States: Lowland Terrestrial Communities*; Martin, W. H.; Boyce, S.; Echternacht, K. Eds.; John Wiley & Sons: New York, **1993,** p.165–198.

Giosan, L.; Syvitski, J.; Constantinescu, S.; Day, J. Climate change: protect the world's deltas. *Nature,* **2014,** *516,* 31–33.

Giri, C.; Ochieng, E.; Tieszen, L. L.; Zhu, Z.; Singh, A.; Loveland, T.; Masek, J.; Duke, N. Status and distribution of mangrove forests of the world using earth observation satellite data. *Global Ecol. Biogeogr.* **2011,** *20,* 154–159.

Glaser, M. Interrelations between mangroves ecosystem, local economy and social sustainability in Caete Estuary. North Braz. *Wetl. Ecol. Manage.* **2003,** *11,* 265–272.

Global Biodiversity Information Facility, http://www.gbif.org (accessed on 11 Jul 2018).

Godoy, M. D. P.; lacerda, L. D. de.Mangroves Response to Climate Change: A Review of Recent Findings on Mangrove Extension and Distribution. *Anais da Academia Brasileira de Ciências,* **2015,** *87, 2,* 651–667.

Goutham, B. M. P.; Roy, S. D.; Krishnan, P.; Kaliyamoorthy, M.; Immanuel, T. Species diversity and distribution of mangroves in Andaman and Nicobar Islands. *Botanica Marina.* **2014,** *57,* 421–432.

Gregory, J. M.; Dixon, K. W.; Stouffer, R. J.; Weaver, A. J.; Driesschaert, E.; Eby, M. A model intercomparison of changes in the Atlantic thermohaline circulation in response to increasing atmospheric CO_2 concentration. *Geophys. Res. Lett.* **2005,** *32,* L12703.

Griffith, W. *Notul.* **1854,** *4,* 663.

Griffith, W. *Trans. Med. Soc. Calcutta,* **1836,** *8,* 10.

Hamilton, L. S.; Snedaker, S. C.; Eds. *Handbook for Mangroves Area Management.* IUCN/Unesco/UNEP. East-West Centre, Honolulu, Hawaii, 1984.

Hanebuth, T. J. J.; Kudrass, H. R.; Linstadter, J.; Islam, B.; Zander, A. M. Rapid coastal subsidence in the central Ganges-Brahmaputra Delta (Bangladesh) since the 17th century deduced from submerged salt-producing kilns. *Geology.* **2013,** *41,* 987- 90.

Harty, C. Planning strategies for mangrove and saltmarsh changes in Southeast Australia. Coastal Management. **2004,** *32,* 405–415.

Hawley, R. C.; Smith, D. M.*The Practice of Silviculture.*6th edition. John Wiley and Sons: New York, 1954.

Hay, C. C.; Morrow, E.; Kopp, R. E.; Mitrovica, J. X. Probabilistic reanalysis of twentieth century sea-level rise. *Nature.* **2015,** *517,* **7535,** 481–484.

Houghton, J.; Ding, Y.; Griggs, D.; Noguer, M.; van der Linden, P.; Dai, X.; Maskell, K.; Johnson, C. Climate Change 2001: The Scientific Basis (Published for the Intergovernmental Panel on Climate Change). Cambridge University Press: Cambridge, United Kingdom, and New York, NY, USA, 2001.

Hull, K. Ancient mangroves reveal rapid sea-level rise. *Australasian Science.* **2005,** *26,* 31–33.

Hutchings P.; Saenger, P. Ecology of mangroves. University of Queensland Press: Queensland (Australia), 1987.

IPCC. Climate Change 2013: The Physical Science Basis. Contribution of Working Group I to the Fifth Assessment Report of the Intergovernmental Panel on Climate Change; Stocker, T. F.; Qin, D.; Plattner, G.-K.; Tignor, M.; Allen, S. K.; Boschung, J.; Nauels, A.; Xia, Y.; Bex, V.; Midgley, P. M., (Eds.). Cambridge University Press: Cambridge, United Kingdom and New York, NY, USA, 2013.

IPCC. Fourth Assessment Report: Climate Change 2007,Working Group I: The Physical Science Basis. 2007. https://www.ipcc.ch/publications_and_data/ar4/wg1/en/ch5s5-5-2.html.

IUCN. *IUCN Red List of Threatened Species.* 2008. http://www.iucnredlist.org.

IUCN. The impact of climatic change and sea level rise on ecosystems. Report for the Commonwealth Secretariat, London, 1989.

Janssonius, H. H. The vessels in the wood of Javan mangrove trees. *Blumea.* **1950,** *6,* 464–469.

Jayatissa, L. P.; Dahdouh-Guebas, F.; Koedam, N. A review of the floral composition and distribution of mangroves in Sri Lanka. *Bot. J. Linn. Soc.* **2002,** *138,* 29–43.

Kathiresan, K. *Rhizophora × annamalayana,* a new species of mangrove. *Environment and Ecology.* **1995,** 240–241.

Kathiresan, K.; Bingham, B. L. Biology of mangroves and mangrove ecosystems. *Advances in Marine Biology.* **2001,** *40,* 81–251.

Kathiresan, K.; Rajendran, N. Coastal mangrove forests mitigated tsunami. *Estuar. Coast. Shelf Sci.* **2005,** *65,* 601–606.

Kiehl, J. T.; Trenberth, K. E. Earth's annual global mean energy budget. *Bull. Amer. Meteorol. Soc.* **1997,** *78,* 197–208

Kim, J.-H.; Dupont, L.; Behling, H.; Versteegh, G. J.M. Impacts of rapid sea-level rise on mangrove deposit erosion: application of taraxerol and Rhizophora records. *Journal of Quaternary Science.* **2005,** *20,* 221–225.

Kirwan, M. L.; Megonigal, J. P. Tidal wetland stability in the face of human impacts and sea-level rise. *Nature,* **2013,** *504,* 53–60.

Klein, R. J. T.; Nicholls, R. J.; Ragoonaden, S.; Capobianco, M.; Aston, J.; E. N. Buckley. Technological options for adaptation to climate change in coastal zones. *Journal of Coastal Research.* **2001,** *17*(3), 531–543.

Knutson, T. R.; Tuleya, R. E. Increased hurricane intensities with CO_2 induced warming assimulated using the GFDL hurricane prediction system. *Clim Dyn.* **1999,** *15,* 503–519.

Kovacs, J. M. Assessing mangrove use at the local scale. *Landsc. Urban Plann.* **1999,** *43,* 201–208.

Krauss, K. W.; Allen, J. A.; Cahoon, D. R. Differential rates of vertical accretion and elevation change among aerial root types in Micronesian mangrove forests. *Estuarine, Coastal and Shelf Science.* **2003,** 56, 251–259.

Krauss, K. W.; Lovelock, C. E.; McKee, K. L.; Lo´pez-Hoffman, L.; Ewe, S. M. L.; Sousa, W. P. Environmental drivers in mangrove establishment and early development: A review. *Aquat. Bot.* **2008,** *89,* 105–127.

Krauss, K. W.; McKee, K. L.; Lovelock, C. E.; Cahoon, D. R.; Saintilan, N.; Reef, R.; Chen, L.How mangrove forests adjust to rising sea level.*New Phytol.* **2014,** Apr; 202, 1, 19–34.

Kristensen, E.; Bouillon, S.; Dittmar, T.; Marchand, C. Organic carbon dynamics in mangrove ecosystems: A review. *Aquat. Bot.* **2008,** *89,* 201–219.

Kumar, R.; Singh, R. D.; Sharma, K. D. Water Resources of India. *Current Science.* **2005,** *89, 5,* 794–811.

Kunstadter, P.; Bird, E. C. F.; Sabhasri, S. Man in the Mangroves. United Nations University: Tokyo, **1986,** pp. 79–86.

Lacerda, L. D.; Conde, J. E.; Bacon, P. R.; Alarcon, C.; D'Croz, L.; Kjerfve, B.; Polania, J.;Vanucci, M. *Mangroves ecosystems in Latin America and the Caribbean: a summary.* In *Conservation and Sustainable Utilization of Mangroves Forests in Latin America and African Regions(Part 1: Latin America). Mangroves Ecosystem Technical Reports 2.* Lacerda, L. D., (Ed.). International Society for Mangroves Ecosystems and International Tropical Timber Organization: Tokyo, 1993.

Lakshmi, M.; Parani, M.; Senthilkumar, P.; Parida, A. Molecular phylogeny of mangroves VIII, Analysis of mitochondrial DNA variation for species identification and relationship in Indian mangrove Rhizophoraceae. *Wetland Ecology and Management.* **2002,** *10,* 355–362.

Lamarck, J.-B. P.A. de M. de; Poiret, J. L. M.; Le palatuvier;Bruguiera Lam. Encyclopédie Méthodique, Botanique, **1798,** *4*(2), 696.

Lang'at, J. K. S.; Kairo, J. G.; Mencuccini, M.; Bouillon, S.; Skov, M. W.; Waldron, S.; Huxham, M. Rapid Losses of Surface Elevation following Tree Girdling and Cutting in Tropical Mangroves, *PLoS ONE,* **2014,** *9,* e107868.

Lessa, G.; Masselink, G. Evidence of a mid-Holocene sea-level highstand from the sedimentary record of a macrotidal barrier and paleoestuary system in northwestern Australia. *Journal of Coastal Research.* **2006,** *22,* 100 – 112.

Lewis III, R. R. Ecological engineering for successful management and restoration of mangrove forests. *Ecol. Eng.* **2005,** *24,* 403–418.

Lewis, R. R. *Creation and restoration of coastal wetlands in Puerto Rico and the U. S. virgin Islands.* In *Wetlands creation and restoration: The status of the science*; Kusler, J. A.; Kentula M. E., (Eds.). Island Press: Washington, D.C., USA, 1990.

Lewis, R. R.; Marshall, M. J. Principles of successful restoration of shrimp aquaculture ponds back to mangrove forests. Programa/resumes de Marcuba '97, September 15/20, Palacio de Convenciones de La Habana, Cuba, 1997.

Lewis, R. R.; Streever, W. Restoration of Mangrove Habitat. Tech Note ERDC TNWRP-VN-RS-3.2. Vicksburg: U. S. Army, Corps of Engineers, Waterways Experiment Station, 2000.

Li, M. S.; Lee, S. Y. Mangroves of China: a brief review. *Forest Ecology and Management.* **1997,** 96: 241–259.

Lopez-Hoffman, L.; Monroe, I. E.; Narvaez, E.; Martinez-Ramos, M.; Ackerly, D. D. Sustainability of mangrove harvesting: how do harvesters' perceptions differ from ecological analysis? *Ecol. Soc.* **2006,** *11*(2) art 14.

Lovelock, C. E.; Cahoon, D. R.; Friess, D. A.; Guntenspergen, G. R.; Krauss, K. W.; Reef, R.; Rogers, K.; Saunders, M. L.; Sidik, F.; Swales, A.; Saintilan, N.; Thuyen, L. X.; Triet, T. The vulnerability of Indo-Pacific mangrove forest to sea-level rise. *Nature, 2015, 526,* 559–563.

Lovelock, C. E.; Ellison, J. C. *Vulnerability of mangroves and tidal wetlands of the Great Barrier Reef to climate change.* In *Climate Change and the Great Barrier Reef: A Vulnerability Assessment*; Johnson, J. E.; Marshall, P. A. Eds.; Great Barrier Reef Marine Park Authority and Australian Greenhouse Office: Australia, 2007; pp. 237–269.

Lovelock, C. E.; Feller, I. C.; Reef, R.; Hickey, S.; Ball, M. C. Mangrove dieback during fluctuating sea levels, *Scientific Reports,* **2017,** *7,* 1680. DOI:10.1038/s41598-017-01927-6.

Lovelock, C. E.; Sorrell, B.; Hancock, N.; Hua, Q.; Swales, A. Mangrove forest and soil development on a rapidly accreting shore in New Zealand. *Ecosystems.* **2010,** *13,* 437–51.

Lugo, A. E.; Snedaker, S. C. The ecology of mangroves. *Annual Review of Ecology and Systematics.* **1974,** *5,* 39–64.

Luther, D.; Greenburg, R. Mangroves: a global perspective on the evolution and conservation of their terrestrial vertebrates. *Bioscience.* **2009,** *59,* 602–612.

Mabberley, D. J. William Theobald (1829–1908): Unwitting Reformer of Botanical Nomenclature? *Taxon,* **1985,** *34,* 152–156.

Macintosh, D. J.; Ashton, E. C. Principles for a code of conduct for the management and sustainable use of mangrove ecosystems. The World Bank: Washington, D.C., 2004.

MacNae, W. A general account of the fauna and flora of mangrove swamps and forests in the Indo-West Pacific region. *Adv. Mar. Biol.* **1968,** *6,* 73–270.

Mandal, R. N.; Das, C. S.; Naskar, K. R. Dwindling Indian Sundarban mangroves: The way out, *Science & Culture,* **2010,** *76* (7–8), 247–254.

Mandal, R. N.; Naskar, K. R. Biodiversity and classification of Indian Mangroves: a review. *Trop. Ecol.* **2008,** *49,* 131–146.

Manson, R. A.; Loneragan, N. R.; Skilleter, G. A.; Phinn, S. R. An evaluation of the evidence for linkages between mangroves and fisheries: a synthesis of the literature and identification of research directions. Oceanography and Marine Biology: An Annual Review, **2005,** *43,* 483–513.

Matthes, H.; Kapetsky, J. M. Worldwide Compendium of Mangrove-associated Aquatic Species of Economic Importance. FAO Fishery Circular No. 814, FAO, Rome, 1988.

Mazda, Y.; Wolanski, E.; Ridd, P. V. The Role of Physical Processes in Mangrove Environments: Manual for the Preservation and Utilization of Mangrove Ecosystems. Terrapub: Tokyo, 2007.

McKee, K. L.; Cahoon, D. R.; Feller, I. Caribbean mangroves adjust to rising sea level through biotic controls on change in soil elevation. *Global Ecol. Biogeogr.* **2007,** *16,* 545–556.

McKee, K. Soil physiochemical patterns and mangrove species distribution-reciprocal effects? *J. Ecol.* **1993,** *81,* 477–487.

McLeod, E.; Salm, R. V. *Managing Mangroves for Resilience to Climate Change.* IUCN, Gland, Switzerland. 2006; p. 64.

Millenium Ecosystem Assessment. Ecosystems and human well-being: Synthesis. Island Press : Washington (DC), 2005.

Milliman, J. D.; Farnsworth, K. L. River Discharge to the Coastal Ocean: A Global Synthesis; Cambridge Univ. Press: 2011.

Mumby, P. J.; Edwards, A. J.; Arias-Gonzalez, J. E.; Lindeman, K. C.; Blackwell, P. G.; Gall, A.; Gorczynska, M. I.; Harbornel, A. R.; Pescod, C. L.; Renken, H.; Wabnitz, C. C. C.; Llewellyn, G. Mangroves enhance the biomass of coral reef fish communities in the Caribbean. *Nature.* **2004,** *427,* 533–536.

Nagelkerken, I.; Blaber, S. J. M.; Bouillon, S.; Green, P.; Haywood, M.; Kirton, L. G.; Meynecke, J.-O.; Pawlik, J.; Penrose, H. M.; Sasekumar, A.; Somer- field, P. J. The habitat function of mangroves for terrestrial and marine fauna: A review. *Aquat. Bot.* **2008,** *89,* 155–185.

Nagelkerken, I.; Roberts, C. M.; van der Velde, G.; Dorenbosch, M.; van Riel, M. C.; Cocheret de la Moriniere, E.; Nienhuis, P. H. How important are mangroves and seagrass beds for coral-reef fish? The nursery hypothesis tested on an island scale. *Mar. Ecol. Progr. Ser.* **2002,** *244,* 299–305.

Naidoo, G. Effects of flooding on leaf water potential and stomatal resistance in *Bruguiera gymnorrhiza* (L.) Lam. *New Phytol.* **1983,** *93,* 369–376.

Naidoo, G. Effects of nitrate, ammonium and salinity on growth of the mangrove *Bruguiera gymnorrhiza* (L.) Lam. *Aquat. Bot.* **1990,** *38,* 209–219.

Naidoo, G. Effects of waterlogging and salinity on plant–water relations and on the accumulation of solutes in three mangrove species. *Aquat. Bot.* **1985,** *22,* 133–143.

Naskar, K. R.; Mandal, R. N. *Ecology and Biodiversity of Indian Mangroves,* vol. I & II, Daya Publishing House: New Delhi, 1999.

Naskar, S.; Palit, P. K. Anatomical and physiological adaptations of mangroves. *Wetlands Ecology and Management.* **2015,** 23 (3), 357–370.

Nathanael, W. R.N. Coconut shells as industrial raw material. Coconut Bulletin; 1964.

Naylor, T.; Totten, E. J.; Jeffries, R. D.; Pozzo, M.; Devey, C. R.; Thompson, S. A. Optimal photometry for color-magnitude diagrams and its application to NGC **2547,** Monthly Notices of the Royal Astronomical Society, **2002,** 335 (2) pp. 291–310.

Nerem, R. S.; Chambers, D.; Choe, C.; Mitchum, G. T. Estimating Mean Sea Level Change from the TOPEX and Jason Altimeter Missions. *Mar. Geod.* **2010,** *33,* 435–446, doi:10.1080/01490419.2010.4910.

Nicholls, R. J.; Cazenave, A. Sea level rise and its impacts on coastal zones. *Science,* **2010,** *328,* 1517–1520.

Nichols, R.; Hoozemans, F.; Marchand, M. Increasing flood risk and wetland losses due to sea-level rise: regional and global analyses. *Global Environ. Change.* **1999,** *9,* S69–S87.

Ning, Z. H.; Turner, R. E.; Doyle, T.; Abdollahi, K. K. *Integrated Assessment of the Climate Change Impacts on the Gulf Coast Region.* Gulf Coast Climate Change Assessment Council (GCRCC) and Louisiana State University (LSU) Graphic Services; 2003.

Nystrom, M.; Folke, C. Spatial resilience of coral reefs. *Ecosystems.* **2001,** *4,* 406 – 417.

Odum, E. P. *Ecology and Our Endangered Life-Support Systems.* Sinauer Associates Inc.; Sunderland, USA; 1989.

Odum, E. P.; Barrett, G. W. *Fundamentals of Ecology,* 5th ed; Brooks-Cole: Belmont, CA; 2004.

Ogden, J. C. *Ecosystem interactions in the tropical coastal seascape*. In *Life and Death of Coral Reefs*; Birkeland, C.; Ed. Chapman & Hall: London, **1997,** pp. 288–297.

Ong, J. E. Mangroves – a carbon source and sink. *Chemosphere.* **1993,** *27,* 1097–1107.

Panda, G. K.; Mishra, M. Emerging trends in environment and sustainable development – the journey from Copenhagen to Paris. In *Environmental Degradation, Prevention, Conservation and Management in India*; Samanta, R. K.; Das, C. S., (Eds.). Progressive Publishers: Kolkata, 2017; pp. 7–27.

Panshin, A. J. An anatomical study of woods of the Philippine mangrove swamps *Philippine J. Sci.*1932, *48,* 143–208.

Pernetta, J. C. Mangrove forests, climate change and sea-level rise: hydrological influences on community structure and survival, with examples from the Indo-West Pacific. A Marine Conservation and Development Report. IUCN, Gland, Switzerland; 1993.

Perry, C. Tropical coastal environments: coral reefs and mangroves. In *Environmental Sedimentology*; Perry, C.; Taylor, K., (Eds.). Blackwell: Oxford, 2007; p. 302–350.

Piou, C.; Feller, I. C.; Berger, U.; Chi, F. Zonation patterns of Belizean offshore mangrove forests 41 years after a catastrophic hurricane. *Biotropica.* **2006,** *38,* 365–372.

Plaziat, J.-C. *Modern and fossil mangroves and mangals: their climatic and biogeographic variability.* In *Marine Palaeoenvironmental Analysis from Fossils*; Bosence, D. W. J.; Allison, P. A., (Eds.). Geological Society: Special Publication No. 83,1995;pp. 73–96.

Polidoro, B. A.; Carpenter, K. E.; Collins, L.; Duke, N. C.; Ellison, A. E.; Ellison, J. C.; Farnsworth, E. J.; Fernando, E. S.; Kathiresan, K.; Koedam, N. E.; Livingstone, S. R.; Miyagi, T.; Moore, G. E.; Nam, V. N.; Ong, J. E.; Primavera, J. H.; Salmo III, S. G.; Sanciangco, J. C.; Sukardjo, S.; Wang, Y.; Yong, J. W. H. The loss of species: mangrove extinction risk and geographic areas of global concern. *PLoS ONE.* **2010,** *5,* 1–10.

Popp, M. Salt resistance in herbaceous halophytes and mangroves *Progress in Botany*1995, *56,* 416–429.

Popp, M.; Polania, J.; Weiper, M. *Physiological adaptations to different salinity levels in mangrove.* In *Towards the rational use of high salinity tolerant plants;* Lieth, H.; Masoom Al A., (Eds.). Vol. 1. Kluwer Academic Publishers: Utrecht, 1993.

Pramanik, M. K. Assessment of the Impacts of Sea Level Rise on Mangrove Dynamics in the Indian Part of Sundarbans Using Geospatial Techniques. *J Biodivers Biopros Dev.* **2015,** *3,* 155. doi:10.4172/2376–0214.1000155 (2015)

Primavera, J. H. Mangroves as nurseries: shrimp populations in mangrove and non-mangrove habitats. *Estuar. Coast. Shelf Sci.* **1998,** *46,* 457–464.

Primavera, J. H. Socioeconomic impacts of shrimp culture. *Aquacult. Res.* **1997,** *28,* 815–827.

Primavera, J. H.; Sadaba, R. B.; Lebata, M. J. H. L.; Altamirano, J. P. *Handbook of Mangroves in the Philippines Panay.* SEAFDEC Aquaculture Department (Philippines) and UNESCO Man and the Biosphere ASPACO Project,2004;p. 106.

Proisy, C.; Gratiot, N.; Anthony, E. J.; Gardel, A.; Fromard, F.; Heuret, P. Mud bank colonization by opportunistic mangroves: A case study from French Guiana using lidar data. *Continental Shelf Research*, **2009,** *29,* 632–641.

Rabinowitz, D. Dispersal properties of mangrove propagules. *Biotropica.*1978, *10,* 47–57.

Ragavan, P.; Jayaraj, R. S. C.; Saxena, A.; Mohan, P. M.; Rvichandran, K. Taxonomical identity of *Rhizophora × annamalayana* Kathir and *Rhizophora × lamarckii* Montrouz (Rhizophoraceae) in the Andaman and Nicobar Islands, India. *Tiawania.* **2015,** *60*(4), 183–193.

Rahman, M. S.; Hossain, G. M.; Khan, S. A.; Uddin, S. N. An annonated checklist of the vascular plants of Sundarban mangrove forest of Bangladesh. *Bangladesh Journal Plant Taxon.* **2015,** *22* (1), 17–41.

Ramsar Secretariat, Wetland Values and Functions: Climate Change Mitigation. Gland, Switzerland, 2001.

Rasolofo, M. V. Use of mangroves by traditional fishermen in Madagascar. *Mangroves and Salt Marshes,* **1997,** *1,* 243–253.

Rasolofoharinoro, M.; Blasco, F.; Bellan, M. F.; Aizpuru, M.; Gauquelin, T.; Denis, J. A remote sensing based methodology for mangrove studies in Madagascar. *International Journal of Remote Sensing.* **1998,** *19*(10), 1873–1886.

Ricklefs, R. E.; Latham, R. E. Global patterns of diversity in mangrove floras. In *Species diversity in ecological communities. Historical and geographical perspective*; Ricklefs, R. E.; Schluter, D., (Eds.), University of Chicago Press: Chicago, **1993,** p. 215–33.

Robertson, A. I.; Duke, N. C. Recruitment, growth and residence time of fishes in a tropical Australian mangrove system. *Estuar. Coast. Shelf Sci.* **1990,** *31,* 723–743.

Rogers, K.; Saintilan, N.; Cahoon, D. R. Surface elevation dynamics in a regenerating mangrove forest at Homebush Bay, Australia. *Wetl. Ecol. Manage.* **2005,** *13,* 587–598.

Rogers, K.; Saintilan, N.; Heijnis, H. Mangrove encroachment of salt marsh in Western Port Bay, Victoria: the role of sedimentation, subsidence, and sea level rise. *Estuaries.* **2005,** *28,* 551–559.

Rollet, B. *Bibliography on mangrove research* 1600–1975. UNESCO, U.K, **1981,** p. 479.

Rönnbäck, P.; Crona, B.; Ingwall, L. The return of ecosystem goods and services in replanted mangrove forests perspectives from local communities in Gazi Bay, Kenya. *Environ. Conserv.* **2007,** *34,* 313–324.

Roth, L. C. *Implications of periodic hurricane disturbance for the sustainable management of Caribbean mangroves.* In *Mangrove ecosystem studies in Latin America and Africa.* Kjerfve, B.; Lacerda, L. D.; Diop, E. H. S., (Eds.). UNESCO: Paris France, 1997.

Roxburgh, W. *Hortus Bengalensis;* Royal Botanical Garden: Calcutta, Serampore, 1814.

Saad, S.; Husain, M. L.; Asano, T. Sediment accretion of a tropical estuarine mangrove: Kemarnan, Terengganu, Malaysia. *Tropics.* **1999,** *8,* 257–266.

Saenger, P. *Mangrove ecology, silviculture and conservation.* Springer-Science +Business media, B. V.; 2002.

Saenger, P. Mangrove vegetation: an evolutionary perspective. *Mar. Freshwater Res.* **1998,** *49,* 277–286.

Saenger, P. *Morphological, anatomical, and reproductive adaptations of Australian mangroves.* In *Mangrove ecosystems in Australia*; Clough, B. F., (Ed.). Australian National University Press: Canberra, 1982.

Saenger, P.; Hegerl, E. J.; Davie, J. D. S., (Eds.). *Global status of mangrove ecosystems.* The Environmentalists 3 (Suppl.), **1983,** pp. 1–88.

Saenger, P.; Hegerl, E. J.; Davie, J. D.S. Global status of mangrove ecosystems. Commission on Ecology Papers No.3. IUCN. Gland, Switzerland, **1983,** p. 88.

Saenger, P.; Moverly, J. Vegetative phenology of mangroves along the Queensland coastline. *Proc. Ecol. Soc. Aust.* **1985,** *13,* 257–265.

Sahu, S. K.; Sing, R.; Kathiresan, K. Deciphering the taxonomical controversies of *Rhizophora* hybrids using AFLP, plastid and nuclear markers. *Aquatic Botany.* **2015,** *125,* 48–56.

Saintilan, N.; Wilson, N.; Rogers, K.; Rajkaran, A.; Krauss, K. W. Mangrove expansion and salt marsh decline at mangrove poleward limits. *Global Change Biology,* **2014,** *20, 1,* 147–157.

Saintilan, N.; Wilton, K. Changes in the distribution of mangroves and saltmarshes in Jervis Bay, Australia. *Wetl. Ecol. Manage.* **2001,** *9,* 409–420.

Salem, M. E.; Mercer, D. E. The economic value of mangroves. *Sustainability.* **2012,** *4,* 359–383.

Sathirathai, S.; Barbier, E. B. Valuing mangroves conservation in southern Thailand. *Contemp. Econ. Policy.* **2001,** *19,* 109–122.

Scholander, P. F. How mangrove desalinate sea water *Physiol. Plant.* **1968,** *21,* 251–261.

Semeniuk, V. Predicting the effect of sea-level rise on mangroves in Northwestern Australia. *Journal of Coastal Research.* **1994,** *10*(4), 1050–1076.

Semesi, A. K. Mangrove management and utilization in eastern Africa. *Ambio.* **1998,** *27,* 620–626.

Serafy, J. E.; Araujo, R. J. Proceedings of the First International Symposium of mangroves as fish habitats, Rosetiel School of Marine and Atmospheric Science, University of Miami, Miami, Florida, 19–21 April,2006,80 Bull Marine Science 3, 2007.

Shackleton, C.; Shackleton, S. The importance of non-timber forest products in rural livelihood security and as safety nets: a review of evidence from South Africa. *South African Journal of Science.* **2004,** 100.

Shaw, J.; Ceman, J. Salt-marsh aggradation in response to late-Holocene sea-level rise at Amherst Point, Nova Scotia, Canada. N 9, **1999,** 439–451.

Shea, E. L.; Dolcemascolo, G.; Anderson, C. L.; Barnston, A.; Guard, C. P.; Hamnett, M. P.; Kubota, S. T.; Lewis, N.; Loschnigg, J.; Meehl, G. Preparing for a changing climate: The potential consequences of climate variability and change. Honolulu: East-West Center, 2001.

Sherman, R. E.; Fahey, T. J.; Martinez, P. Hurricane impacts on a mangrove forest in the Dominican Republic, damage patterns and early recovery. *Biotropica.* **2001,** *33,* 393–408.

Sheue, C-R; Liu, H-Y; Tsai, C-C; Yang, Y-P. Comparison of *Ceriops pseudodecandra* sp. nov. (Rhizophoraceae), a new mangrove species in Australasia, with related species. *Botanical Studies.* **2010,** *51,* 237–248.

Sheue, C. R.; Liu, H. Y.; Tsai, C. C.; Rasid, S. M. A.; Yong, J. W. H.; Yang, Y. P. On the morphology and molecular basis of segregation of *Ceriops zippeliana* and *C. decandra* (Rhizophoraceae) from Asia. *Blumea,* **2009,** *54,* 220–227.

Smith III, T. J. *Forest structure.* In *Tropical Mangrove Ecosystems.* Robertson, A. I.; Alongi, D. M., (Eds.). American Geophysical Union, Washington, D.C. 1992; p. 251–292.

Smith III, T. J.; Robblee, M. B.; Wanless, H. R.; Doyle, T. W. Mangroves, hurricanes, and lightning strikes. *Bioscience.* **1994,** *44,* 256–262.

Snedaker, S. C. Mangroves and climate change in the Florida and Caribbean region: scenariosand hypotheses. *Hydrobiologia,***1995,** *295,* 43–49.

Snedaker, S. C.; Araujo, R. J. Stomatal conductance and gas exchange in four species of Caribbean mangroves exposed to ambient and increased CO_2. *Marine and Freshwater Research.* **1998**, *49,* 325–327.

Snedaker, S. *Impact on mangroves.* In *Climate Change in the Intra-American Seas: Implications of Future Climate Changeon the Ecosystems and Socio-economic Structure of the Marine and CoastalRegimes of the Caribbean Sea, Gulf of Mexico, Bahamas and N. E. Coast ofSouth America;* Maul, G. A. Ed.; Edward Arnold: London,1993;pp. 282–305.

Solomon, S.; Qin, D.; Manning, M.; Alley, R. B.; Berntsen, T.; Bindoff, N. L.; Chen, Z.; Chidthaisong, A.; Gregory, J. M.; Hegerl, G. C.; Heimann, M.; Hewitson, B.; Hoskins, B. J.; Joos, F.; Jouzel, J.; Kattsov, V.; Lohmann, U.; Matsuno, T.; Molina, M.; Nicholls, N.; Overpeck, J.; Raga, G.; Ramaswamy, V.; Ren, J.; Rusticucci, M.; Somerville, R.; Stocker, T. F.; Whetton, P.; Wood, R. A.; Wratt, D. *Technical summary.* In *Climate Change 2007: The Physical Science Basis. Contribution of Working Group I to the Fourth Assessment Report of the Inter- governmental Panel on Climate Change;* Solomon, S.; Qin, D.; Manning, M.; Chen, Z.; Marquis, M.; Averyt, K. B.; Tignor, M.; Miller, H. L., (Eds.). Cambridge University Press: Cambridge, United Kingdom and New York, NY, USA, 2007.

Spalding, M.D; Blasco, F.; Fields, C.D, Eds.; *World mangrove atlas.* ISME/ITTO, Okinawa, 1997.

Stafleu, F. A.; Cowan, R. S. Taxonomic literature: A selective guide to botanical publications and collections with dates, commentaries and types. **1985,** Vol. V, p. 90–91.

Stammer, D.; Cazenave, A.; Ponte, R. M.; Tamisiea, M. E. Causes for contemporary regional sea level changes. *Annu. Rev. Mar. Sci.* **2013,** *5,* 1, 21–46.

Steenis, C. G. G. J. The distribution of mangrove plant genera and its significance for palaeogeography *Proc. Kon. Net.; Amsterdam Ser. C.* **1962,** *65,* 164–169.

Sullivan, C. 2005. The importance of mangroves Available: www.vi_shandwildlife.com/ Educaton/FactSheet/PDF_docs/28Mangrrovespdf.; 2009.

Svetlana, J.; Moore, J. C.; Grinsted, A.; Woodworth, P. L.; Recent global sea level acceleration started over 200 years ago?.*Geophysical Research Letters.*April **2008,** *35,* 8. Bibcode:2008GeoRL.35.8715J.doi:10.1029/2008GL033611.

Swales, A.; Bentley, S. J. Sr.; Lovelock, C. E. Mangrove-forest evolution in a sediment-rich estuarine system: opportunists or agents of geomorphic change? *Earth Surf. Process. Landf.* **2015,** *40,* 1672 – 87.

Syvitski, J. P. M.; Kettner, A. J.; Overeem, I.; Hutton, E. W. H.; Hannon, M. T. Sinking deltas due to human activities. *Nat. Geosci.* **2009,** *2,* 681 – 86.

Theobald, W. F. *Mason, Burmah,* 3rd edn, **1883,** *2,* 480.

The Plant List. www.theplantlist.org

Thom, B. G. *Coastal landforms and geomorphic processes.* In *The Mangrove Ecosystem: Research Methods;* Snedaker, S. C.; Snedaker, J. G. Eds.; UNESCO, Paris, 1984; p. 3–15.

Thom, B. G. *Mangrove ecology – a geomorphological perspective.* In *Mangrove ecosystems in Australia;*Clough, B. F., (Ed.). Australian National University Press: Canberra, 1982.

Tomlinson, P. B. *The Botany of mangroves,* Cambridge University Press: Cambridge, 1986.

Trenberth, K. Uncertainty in hurricanes and global warming. *Science.* **2005,** *308,* 1753–1754.

Triest, L. Molecular ecology and biogeography of mangrove trees towards conceptual insights on gene flow and barriers: a review. *Aquatic Botany.* **2008**, *89,* 138–154.

Twilley, R. R. *Properties of mangrove ecosystems and their relation to the energy signature of coastal environments.* In *Maximum power.* Hall, C. A. S. Ed.; University of Colorado Press: Boulder, Colorado, USA, 1995; p. 43–62.

UNESCO. Coastal systems studies and sustainable development. Proceedings of the COMAR Interregional Scientific Conference, UNESCO, Paris, 21–25 May, 1991. UNESCO, Paris, **1992**, p. 276.

United Nations Environment Programme. Assessment and monitoring of climatic change impacts on mangrove ecosystems. UNEP Regional Seas Reports and Studies. Report no. 154, 1994.

Untawale, A. G. *Exploitation of mangroves in India.* In *Mangroves Ecosystems of Asia and the Pacific: Status, Exploitation and Management;* Field, C. D.; Dartnall, A. J., (Eds.). Australian Development Assistance Bureau and Australian Committee for Mangroves Research, Townsville, **1987**, p. 222.

Valiela, I.; Bowen, J.; York, J. Mangrove forests: one of the world's threatened major tropical environments. *Bioscience.* **2001**, *51,* 807–815.

Vannucci, M. Supporting appropriate mangrove management. International News Letter of Coastal Management-Intercoast Network: Special edition 1, **1997**, pp. 1–3.

Vedeld, P.; Angelsen, A.; Sjaastad, E.; Berg, G. K., (Eds.). *Counting on the Environment: Forest Incomes and the Rural Poor.* Environmental Economics Series Paper No. 98. World Bank, Washington, DC, 2004.

Vliet, G J. C. M.; van, Wood anatomy of the Combretaceae. *Blumea.* **1979**, *25,* 141–223.

Wakushima, S.; Kuraishi, S.; Sakurai, N. Soil salinity and pH in Japanese mangrove forests and growth of cultivated mangrove plants in different soil conditions. *J. Plant Res.* **1994**, *107,* 39–46.

Wakushima, S.; Kuraishi, S.; Sakurai, N.; Supappibul, K.; Siripatanadilok, S. Stable soil pH of Thai mangroves in dry and rainy seasons and its relation to zonal distribution of mangroves. *J. Plant Res.* **1994**, *107,* 47–52.

Walker, M. D.; Wahren, C. H.; Hollister, R. D. Plant community responses to experimental warming across the tundra biome. *P. Natl. Acad. Sci. USA* **2006**, *103,* 1342–46.

Walsh, G. E. *Mangroves: a review.* In *Ecology of Halophytes;* Reimhold R. J.; Queen, W. H., (Eds.). Academic Press: New York, 1974.

Walsh, K. J. E.; Ryan, B. F. Tropical cyclone intensity increase near Australia as a result of climate change. *J Climate*, **2000**, *13,* 3029–3036.

Walters, B. B. Ecological effects of small-scale cutting of Philippine mangrove forests. *For. Ecol. Manage* **2005**, *206,* 331–348.

Walters, B. B. Local management of mangrove forests: effective conservation or efficient resource exploitation? *Hum. Ecol.* **2004**, *32,* 177–195.

Walters, B. B.; Rönnbäck, P.; Kovacs, J. M.; Crona, B.; Hussain, S. A.; Badola, R.; Primavera, J. H.; Barbier, E.; Dahdouh-Guebas, F. Ethnobiology, socio-economics and management of mangrove forests: A review. *Aquat. Bot.* **2008**, *89,* 220–236.

Walton, M. E.; LeVay, L.; Lebata, J. H.; Binas, J.; Primavera, J. H. Seasonal abundance, distribution and recruitment of mud crabs (*Scylla* spp.) in replanted mangroves. *Estuar. Coast. Shelf Sci.* **2006**, *66,* 493–500.

Wang, W.; Yan, Z.; You, S.; Zhang, Y.; Chan, L.; Lin, G. *Tree.* **2011**, *25,* 953–963.

Ward, R. D.; Friess, D. A.; Day, R. H.; MacKenzie, R. A. Impacts of climate change on mangrove ecosystems: a region by region overview. Ecosystem Health and Sustainability, **2016,** *2,* 4, e01211. doi:10.1002/ehs2.1211.

Webb, E. L.; Friess, D. A.; Krauss, K. W.; Cahoon, D. R.; Guntenspergen, G. R.; Jacob P. (2013) A global standard for monitoring coastal wetland vulnerability to accelerated sea-level rise. *Nature Climate Change,* **2013,** vol. 3. May.

Weiper, M. Physiological and structural studies on salt regulation in mangroves. Unpublished PhD Dissertation. Wiley & Sons Inc.: Westphalian Wilhelms University, Münster, 1995.

Wells, S.; Ravilous, C.; Corcoran, E. In the front line: Shoreline protection and other ecosystem services from mangroves and coral reefs. United Nations Environment Programme World Conservation Monitoring Centre: Cambridge, UK, **2006,** p. 33.

Whelan, K. R. T.; Smith, T. J.; Cahoon, D. R.; Lynch, J. C.; Anderson, G. H. Groundwater control of mangrove surface elevation: Shrink and swell varies with soil depth. *Estuaries.* **2005,** *28*(6), 833–843.

Wilson, R. Impacts of Climate Change on Mangrove Ecosystems in the Coastal and Marine Environments of Caribbean Small Island Developing States (SIDS) Caribbean marine climate change report card. *Science Review,* **2017,** 60–82.

Winterwerp, J. C.; Erftemeijer, P. L. A.; Suryadiputra, N.; Eijk, P. van.; Zhang, L. Defining Eco-Morphodynamic Requirements for Rehabilitating Eroding Mangrove-Mud Coasts. *Wetlands,* **2013,** *33,* 515–526.

Wolanski, E. Transport of sediment in mangrove swamps. *Hydrobiologia.* **1995,** *295,* 31–42.

Woodroffe, C. *Coasts: Form, Process and Evolution.* Cambridge University Press: Cambridge, UK, 2002.

Woodroffe, C. D. Mangrove sediments and geomorphology. In *Tropical Mangrove Ecosystems*; Robertson, A. I.; Alongi, D. M., (Eds.). American Geophysical Union: Washington, D.C., 1992; pp. 7–41.

Woodroffe, C. D. Pacific island mangroves: distributions and environmental settings. *Pacific Sci.* **1987,** *41,* 166–185.

Woodroffe, C. D. Response of tide-dominated mangrove shorelines in northern Australia to anticipated sea-level rise. *Earth Surface Processes and Landforms.* **1995,** *20*(1), 65–85.

Woodroffe, C. D. The impact of sea-level rise on mangrove shoreline; *Progress in Physical Geography,***1990,** *14,* 483–502.

Woodroffe, C. D.; Grime, D. Storm impact and evolution of a mangrove fringed chenier plain, Shoal Bay, Darwin, Australia. *Mar. Geol.* **1999,** *159,* 303–321.

Woodroffe, C. D.; Grindrod, J. Mangrove biogeography: the role of quaternary environmental and sea-level change. *Journal of Biogeography.* **1991,** *18,* 479–492.

Woodroffe, C. D.; Rogers, K.; McKee, K. L.; Lovelock, C. E.; Mendelssohn, I. A.; Saintilan, N. Mangrove Sedimentation and Response to Relative Sea-Level Rise. *Annu. Rev. Mar. Sci.* **2016,** *8,* 243–66.

Woodroffe, C. D.; Thom, B. G.; Chappell, J. Development of widespread mangrove swamps in mid-Holocene times in northern Australia. *Nature,* **1985,** *317,* 711–713.

Woodruff, J. D.; Irish, J. L.; Camargo, S. J. Coastal flooding by tropical cyclones and sea-level rise. *Nature.* **2013,** *504,* 44–52.

Yang, H. J.; Wu, M. Y.;Liu, W. X.Community structure and composition in response to climate change in a temperate steppe.*Glob Change Biol.* **2011,** *17,* 452–65.

Yulianto, E.; Rahardjo, A. T.; Noeradi, D.; Siregar, D. A.; Hirakawa, K. A Holocene pollen record of vegetation and coastal environmental changes in the coastal swamp forest at Batulicin, south Kalimantan, Indonesia. *Journal of Asian Earth Sciences.* **2005,** *25,* 1–8.

Zhang, R. T.; Lin, P. Studies on the flora of mangrove plants from the coast of China. *J. Xiamen Univ. (Nat. Sci.)* **1984,** *23,* 232–239, in Chinese, with English abstract.

GLOSSARY

Abaxial surface: Lower surface of the leaf.

Accretion: It is here to refer to the process of new layers being slowly added to old ones through sediments deposition.

Adaxial surface: Upper surface of the leaf.

Allochthonous: It refers to the rocks or deposits formed in a place other than where they are found, i.e., it originates from outside a system.

Anatomy: The branch of biology is dealing with the internal structures of organisms.

Anthropogenic: It is related to the activities of humans on nature

Aqueous tissue: The tissue contains parenchymatous cells in which water remain stored.

Biomass: It refers to the total mass of all living organisms or of a particular set of organisms in an ecosystem.

Buffer zone: It is considered to be an area situated between land and sea, which act to reduce adverse effects arising due to climate change and sea level rise.

Canopy: A coverage formed by the leafy upper branches of the trees in a forest.

Catchment: The area is connected with drainage system of a river or of water body.

Channelization: It refers to the system to be connected with rivers or streams by means of an artificial channel.

Climate change: It refers to any long-term change in Earth's climate, including warming, cooling and changes besides temperature.

Climber: A climber may be defined as a plant with weak stem growing beside any support and climbing up.

Colonization: It refers here to the process by which population of any biological entity to occupy new area for their establishment.

Community: Community is defined as a group of populations of different species in a given area.

Conservation: It refers to management through which biological diversity as a whole is protected, restored and checked from the further loss.

Creeper: A plant that grows and spreads upon the ground or surface.

Critically endangered (CR): A taxon is Critically Endangered when the best available evidence indicates that it has undergone a significant decline in the near past, or is projected to experience a significant decline in the near future and is therefore considered to be facing an extremely high risk of extinction in the wild.

Data deficient (DD): A taxon is Data Deficient when there is inadequate information to make a direct, or indirect, assessment of its risk of extinction based on its distribution and/or population status.

Dorsiventral leaf: Leaf showing clear differentiation of dorsal and ventral sides.

Elevation: It is here to refer to the height of a point above (or below)sea level or mean*sea level*, as usually referred to.

Endangered (EN): A taxon is Endangered when the best available evidence indicates that it has shown a considerable population loss in the near past, or is projected to experience a significant decline in the near future and it is therefore considered to be facing a very high risk of extinction in the wild.

Epiphyte: A plant that grows on the other plant, deriving nutrients and water it or from air, rain, etc.

Estuary: The wide part of a river where it flows into the sea.

Fern: A plant with delicate leaves, but no flowers

Flora: Listing of species in a given area is considered as flora.

Geographic Information System (GIS): It comprises an organized collection of computer hardware, software, geographic data, and personnel designed to capture, store, update, manipulate, analyze, and display all forms of geographical information.

Geomorphology: It deals with the study of landforms on a planet's surface.

Global warming: It refers to the long-term increase in Earth's average temperature.

Grass: A grass may be defined as monocot plant with fibrous roots.

Greenbelt: It refers to an area of open land used for particular activities, including plantation of mangroves seedlings, nurturing of young mangroves and their management.

Habitat: The place in which a particular species live is known as its habitat. The habitat of an organism manifests a set of environmental conditions to which it is adjusted to grow.

Herb: An herb may be defined as a small plant with weak and succulent stem.

Holocene: It refers to the present, post-Pleistocene geologic epoch of the Quaternary period, including the last 10,000 years; the recent or Postglacial period.

Hydrology: It deals with the study of the movement of water from the sea through the air to the land and back to the sea, including the properties, distribution, and circulation of water on or below the Earth's surface and in the atmosphere.

Incipient vivipary: It refers to the condition of seed germination while cotyledon remains hidden within the fruit coat. This condition also refers to crypto vivipary.

Iso-bilateral leaf: Leaf without clear differentiation of dorsal and ventral sides.

Least concern (LC): A taxon is Least Concern when it has been evaluated against the criteria and does not qualify for Critically Endangered, Endangered, Vulnerable or Near Threatened. Widespread and abundant taxa are included in this category.

Lenticel: A small aperture develops on stem surface or in leaf to help breathing of plants.

Lepidote: Leaf surface covered with small scales.

Mesophyte: It refers to the plant communities growing under condition of well balanced moisture supply.

Morphology: The branch of biology is dealing with the forms and structures of organisms.

Msophyll: The tissue containing chlorophyll in their cells is known as mesophyll.

Neap tide: A tide in the sea in which there is a very small difference between the level of the water at 'high tide' and that at 'low tide.'

Near threatened (NT): A taxon is Near Threatened when it has been evaluated against the criteria but does not qualify for Critically Endangered, Endangered or Vulnerable now, but is close to qualifying for or is likely to qualify for a threatened category in the near future.

Palisade tissue: Upper layer of a ground tissue in a leaf, consisting of elongated cells beneath and perpendicular to the upper epidermis and constituting the primary area of photosynthesis.

Palm: A monocot tree growing usually straight with mass of long leaves at the top.

Persistence: It may be referred to constancy of mangroves in habitats over time, regardless of environmental perturbation as defined in view of estuarine ecosystems (Boesch, 1974; Gilman et al., 2008).

Phenology: The science dealing with the particular occurrence of plant and animal life cycle in relation to the influence of climate.

Population: Population is defined as the total number of individuals of the taxon.

Propagule: A living entity is capable of dispersal and of producing a new mature individual (e.g. a spore, seed, fruit, egg, larva, or part of or an entire individual). Gametes and pollen are not considered propagules in this context.

Recruitment: It refers to new members established into a population by reproduction or immigration.

Refugia: It refers to an area that may remain relatively unaltered due to climate change and is inhabited by plants and animals during a period of continental climatic change (e.g., glaciation) and remains as a center of relict forms from which a new dispersion and speciation may occur after climatic readjustment. It is considered to be secure areas that are protected by natural factors and human intervention from a variety of stresses.

Region: A region is defined here to be subglobal geographical area, such as a continent, country, state, or province.

Remote sensing: It deals with the methods for gathering data on a large or landscape scale which do not involve on-the-ground measurement, especially satellite photographs and aerial photographs; often used in conjunction with Geographic Information Systems.

Resilience: It is defined to mean the ability to recover from disturbance to some more or less persistent state (Boesch, 1974; Gilman et al., 2008). Resilience refers to the capacity of a mangrove to naturally migrate landward in response to sea-level rise, and that facilitates mangroves ecosystem to be able to withstand the effects of the stress so as to maintain its functions, processes and structure (Carpenter et al., 2001; Nystrom and Folke, 2001).

Resistance: It is defined to mean mangroves' ability to keep pace with rising sea level without alteration to its functions, processes and structure (Odum, 1989; Bennett et al., 2005; Alongi, 2008).

Rhizomatous axis: The thick stem of a plant that grows along or under the ground and has roots growing from it.

Sclerides: The tissue consists of cells which have rigid cell wall made of majorly lignin properties.

Sea level rise: It refers to an increase in the volume of water in the world's oceans, resulting in an increase in global **Mean Sea Level** (MSL). Two major mechanisms cause **sea level** to **rise**. First, as ocean temperatures-**rise**, the warmer water expands **(thermal expansion of the water in the oceans)**. Second, melting or shrinking land ice, such as mountain glaciers and polar ice sheets, is releasing water into the oceans. Melting of floating Ice shelfs or icebergs at sea does not raise sea levels. Sea level rise at specific locations may be more or less than the global average. Local factors might include tectonic effects, subsidence of the land, tides, currents, storms, etc. (Fischlin et al., 2007).

Sea-level: Itis the average height of the sea/ocean, used as the basis for measuring the height of all places on land (Oxford Advanced learner's Dictionary, 7th edition). Also, it is considered to be the level of the surface of the sea with respect to the land, taken to be the mean level between high and low tide, and used as a standard base for measuring heights and depths (Collins English Dictionary – Complete & Unabridged, 2012, Digital Edition).

Seedling: A young plant that has grown from a seed.

Shrub: A shrub may be defined as a woody plant < 8 m tall with many stems near the base.

Spring tide: A tide in which there is a very great rise and fall of the sea and which happens near the new moon and the full moon each month.

Stability: The condition has been used here to refer to environmental constancy, community persistence, and community or ecosystem response to disturbance, exclusively in perspective of climate change (Gilman et al., 2008).

Surface Elevation Tables (SETs): It is a portable mechanical leveling device for measuring the relative elevation of wetland sediments.

Swamp: An area of ground that is wet or covered with water and in which plants, trees, etc. are growing.

Synonym: Two or more scientific names are used to indicate the same species.

Terrigenous sediment: It refers to the sediment usually derived from erosion of rocks; i.e., from terrestrial environment. Terrigenous sediment is consists of sand, mud and silt carried by rivers to the sea.

Transpiration: It is a physiological process through which the loss of water vapor from a plant to the outside atmosphere occurs.

Tree: A tree may be defined as a woody plant > 5 m tall with a single stem.

Twiner: A twiner may be defined as a plant with weak stem twining round the support.

Vegetation: Vegetation may be defined as the sum total of plant population covering an area; vegetation may comprise one species or a combination of different species in a given area.

Vivipary: It refers to the condition of seed germination while fruit remains attached to the mother plant.

Vulnerable (VU): A taxon is Vulnerable when the best available evidence indicates that it meets any of the criteria A to E for Vulnerable, and it is therefore considered to be facing a high risk of extinction in the wild.

Xerophyte: Plants adapted for growth in dry or physiologically dry condition of habitats are known as xerophytic plants. Morphological attributes of plants that are adapted to sustain in xerophytic condition are known as Xeromorphy.

INDEX

Milton Keynes UK
Ingram Content Group UK Ltd.
UKHW022059141024
449569UK00031B/1708

9 781774 634066